D0845475

Quarks, Leptons and the Big Bang

Quarks, Leptons and the Big Bang

Jonathan Allday

The King's School, Canterbury

Institute of Physics Publishing
Bristol and Philadelphia

© IOP Publishing Ltd 1998

All rights reserved. No part of this publication may be reproduced, stored in a retrieval system or transmitted in any form or by any means, electronic, mechanical, photocopying, recording or otherwise, without the prior permission of the publisher. Multiple copying is permitted in accordance with the terms of licences issued by the Copyright Licensing Agency under the terms of its agreement with the Committee of Vice-Chancellors and Principals.

British Library Cataloguing-in-Publication Data

A catalogue record for this book is available from the British Library.

ISBN 0 7503 0461 8 (hbk)
     0 7503 0462 6 (pbk)

Library of Congress Cataloging-in-Publication Data are available

Published by Institute of Physics Publishing, wholly owned by The Institute of Physics, London

Institute of Physics Publishing, Dirac House, Temple Back, Bristol BS1 6BE, UK

US Editorial Office: Institute of Physics Publishing, The Public Ledger Building, Suite 1035, 150 South Independence Mall West, Philadelphia, PA 19106, USA

Typeset in TEX using the IOP Bookmaker Macros
Printed in the UK by J W Arrowsmith Ltd, Bristol

Contents

Preface

It is difficult to know what to say in the preface to a book. Certainly it should describe what the book is about.

This is a book about particle physics (the strange world of objects and forces that exists at length scales much smaller than the size of an atom) and cosmology (the study of the origin of the universe). It is quite extraordinary that these two extremes of scale can be drawn together in one book. Yet the advances of the past couple of decades have shown that there is an intimate relationship between the world of the very large and the very small. The key moment that started the forging of this relationship was the discovery of the expansion of the universe in the 1920s. If the universe has been expanding since its creation (some 15 billion years ago) then at some time in the past the objects within it were very close together and interacting by the forces that particle physicists study. At one stage in its history the whole universe was the microscopic world. In this book I intend to take the reader on a detailed tour of the microscopic world and then through to the established ideas about the big bang creation of the universe and finally to some of the more recent refinements and problems that have arisen in cosmology. In order to do this we need to discuss the two most important fundamental theories that have been developed this century: relativity and quantum mechanics. The treatment is more technical than a popular book on the subject, but much less technical than a textbook.

Another thing that a preface should do is to explain what the reader is expected to know in advance of starting this book.

In this book I have assumed that the reader has some familiarity with energy, momentum and force at about the level expected of a modern GCSE candidate. I have also assumed a degree of familiarity with mathematics—again at about the modern GCSE level. However, readers who are put off by mathematics can always leave the boxed calculations for another time without disturbing the thread of the argument.

Finally, I guess that the preface should give some clue as to the spirit behind the book. In his book *The Tao of Physics* Fritjof Capra says that physics is a 'path with a heart'. By this he means that it is a way of thinking that can lead to some degree of enlightenment not just about the world in which we live, but also about us, the people who live in it. Physics is a human subject, despite the dry mathematics and formal presentation. It is full of life, human tragedy, exhilaration, wonder and very hard work. Yet by and large these are not words that most people would associate with physics after being exposed to it at school (aside from hard work that is). Increasingly physics is being marginalized as an interest at the same time as it is coming to grips with the most fundamental questions of existence. I hope that some impression of the life behind the subject comes through in this book.

Acknowledgments

I have many people to thank for their help and support during the writing of this book.

Liz Swinbank, Susan Oldcorn and Lewis Ryder for their sympathetic reading of the book, comments on it and encouragement that I was on the right lines.

Professors Brian Foster and Ian Aitchison for their incredibly detailed readings that found mistakes and vagaries in the original manuscript. Thanks to them it is a much better book. Of course any remaining mistakes can only be my responsibility.

Jim Revill, Al Troyano and the production team at Institute of Physics Publishing.

Many students of mine (too many to list) who have read parts of the book. Various Open University students who have been a source of inspiration over the years and a captive audience when ideas that ended up in this book have been put to the test at summer schools.

Graham Farmello, Gareth Jones, Paul Birchley, David Hartley and Becky Parker who worked with Liz and I to spice up A level physics by putting particle physics in.

Finally thanks to family and friends.

Carolyn, Benjamin and Joshua who have been incredibly patient with me and never threatened to set fire to the computer.

My parents Joan and Frank who knew that this was something that I really wanted to do.

John and Margret Gearey for welcoming me in.

Robert James, a very close friend for a very long time.

Richard Houlbrook, you see I said that I would not forget you.

Jonathan Allday
jonathan@jacant.demon.co.uk
November 1997

Prelude

Setting the scene

What is particle physics?

Particle physics attempts to answer some of the most basic questions about the universe:

- are there a small number of different types of objects from which the universe is made?
- do these objects interact with each other and, if so, are there some simple rules that explain what will happen?
- how can we study the creation of the universe in a laboratory?

The topics that particle physicists study from one day to the next have changed as the subject has progressed, but behind this progression the final goal has remained the same—to try to understand how the universe came into being.

Particle physics tries to answer questions about the origin of our universe by studying the objects that are found in it and the ways in which they interact. This is like someone trying to learn how to play chess by studying the shapes of the pieces and the ways in which they move across the board.

Perhaps you think that this is a strange way to try to find out about the origin of the universe. Unfortunately, there is no other way. There are instruction manuals to help you learn how to play chess; there are no instruction manuals supplied with the universe. Despite this handicap an impressive amount has been understood by following this method.

Some people argue that particle physics is fundamental to all the sciences as it strips away the layers of structure that we see in the world and plunges down to the smallest components of matter. This study applies equally to the matter that we see on the earth and that which is in the stars and galaxies that fill the whole universe. The particle physicist assumes that all matter in the universe is fundamentally the same and that it all had a common origin in the big bang that created our universe. (This is a reasonable assumption as we have no evidence to suggest that any region of the universe is made of a different form of matter. Indeed we have positive evidence to suggest the opposite.)

The currently accepted scientific theory is that our universe came into being some fifteen billion years ago in a gigantic explosion. Since then it has been continually growing and cooling down. The matter created in this explosion was subjected to unimaginable temperatures and pressures. As a result of these extreme conditions, reactions took place that were crucial in determining how the universe would turn out. The structure of the universe that we see now was determined just after its creation.

If this is so, then the way that matter is structured now must reflect this common creation. Hence by building enormous and expensive accelerating machines and using them to smash particles together at very high energies, particle physicists can force the basic constituents of matter into situations that were common in the creation of the universe—they produce miniature big bangs. Hardly surprisingly, matter can behave in very strange ways under these circumstances.

Of course, this programme was not worked out in advance. Particle physics was being studied before the big bang theory became generally accepted. However, it did not take long before particle physicists realized that the reactions they were seeing in their accelerators must have been quite common in the early universe. Such experiments are now providing useful information for physicists working on theories of how the universe was created.

In the past twenty years this merging of subjects has helped some huge leaps of understanding to take place. We believe that we have an accurate understanding of the evolution of the universe from the first 10^{-5} seconds onwards (and a pretty good idea of what happened even earlier). By the time you have finished this book, you will have met many of the basic ideas involved.

Why study particle physics?

All of us, at some time, have paused to wonder at our existence. As children we asked our parents embarrassing questions about where we came from (and, in retrospect, probably received some embarrassing answers). In later years we may ask this question in a more mature form, either in accepting or rejecting some form of religion. Scientists that dedicate themselves to pure research have never stopped asking this question.

It is easy to conclude that society does not value such people. Locking oneself away in an academic environment 'not connected with the real world' is generally regarded as a (poorly paid) eccentricity. This is very ironic. Scientists are engaged in studying a world far more real than the abstract shuffling of money on the financial markets. Unfortunately, the creation of wealth and the creation of knowledge do not rank equally in the minds of most people.

Against this background of poor financial and social status it is a wonder that anyone chooses to follow the pure sciences; their motivation must be quite strong. In fact, the basic motivation is remarkably simple.

Everyone has, at some time, experienced the inner glow that comes from solving a puzzle. This can take many forms, such as maintaining a car, producing a difficult recipe, solving a jigsaw puzzle, etc. Scientists are people for whom this feeling is highly magnified. Partly this is because they are up against the ultimate puzzle. As a practising and unrepentant physicist I can testify to the feeling that comes from prising open the door of nature by even a small crack and understanding something new for the first time. When such an understanding is achieved the feeling is one of personal satisfaction, but also an admiration for the puzzle itself. Few of us are privileged enough to get a glimpse through a half-open door, like an Einstein or a Hawking, but we can all look over their shoulders. The works of the truly great scientists are part of our culture and should be treated like any great artistic creation. Such work demands the support of society.

Unfortunately, the appreciation of such work often requires a high degree of technical understanding. This is why science is not valued as much as it might be. The results of scientific experiments are often felt to be beyond the understanding, and hence the interest, of ordinary people. Scientists are to blame. When Archimedes jumped out of his bath and ran through the streets shouting 'Eureka!' he did not stop to explain his actions to the passers by. Little has changed in this respect over the intervening centuries. We occasionally glimpse a scientist running past shouting about some discovery, but are unable to piece anything together from the fragments that we hear. Few scientists are any good at telling stories.

The greatest story that can be told is the story of creation. In the past few decades we have been given an outline of the plot, and perhaps a glimpse of the last page. As in all mystery stories the answer seems so obvious and simple, it is a wonder that we did not think of it earlier. This is a story so profound and wonderful that it must grab the attention of anyone prepared to give it a moment's time.

Once it has grabbed you, questions as to why we should study such things become irrelevant—*it is obvious that we must.*

Chapter 1

The standard model

This chapter is a brief summary of the theories discussed in the rest of this book. The standard model of particle physics—the current state of knowledge about the structure of matter—is described and an introduction provided to the 'big bang' theory of how the universe was created. We shall spend the rest of the book exploring in detail the ideas presented in this chapter.

1.1 The fundamental particles of matter

It is remarkable that a list of the fundamental constituents of matter easily fits on a single piece of paper. It is as if all the recipes of all the chefs that have been and will be could be reduced to combinations of twelve simple ingredients.

The twelve particles from which all forms of matter are made are listed in table 1.1. Twelve particles, that is all that there is to the world of matter.

The twelve particles are divided into two distinct groups called the *quarks* and the *leptons* (at this stage don't worry about where the names come from). Quarks and leptons are distinguished by the different ways in which they react to the fundamental forces.

There are six quarks and six leptons. The six quarks are called up, down, strange, charm, bottom and top[1] (in order of mass). The

Table 1.1 The fundamental particles of matter.

Quarks		Leptons	
up	(u)	electron	(e^-)
down	(d)	electron-neutrino	(ν_e)
strange	(s)	muon	(μ^-)
charm	(c)	muon-neutrino	(ν_μ)
bottom	(b)	tau	(τ^-)
top	(t)	tau-neutrino	(ν_τ)

six leptons are the electron, the electron-neutrino, the muon, muon-neutrino, tau and tau-neutrino. As their names suggest, their properties are linked.

Already in this table there is one familiar thing and one surprise.

The familiar thing is the electron, which is one of the constituents of the atom and the particle that is responsible for the electric current in wires. Electrons are fundamental particles, which means that they are not composed of any smaller particles—they do not have any pieces inside them. All twelve particles in table 1.1 are thought to be fundamental—they are all distinct and there are no pieces within them.

The surprise is that the proton and the neutron are not mentioned in the table. All matter is composed of atoms of which there are 92 naturally occurring types. Every atom is constructed from electrons which orbit round a small, heavy, positively charged nucleus. In turn the nucleus is composed of protons, which have a positive charge, and neutrons, which are uncharged. As the size of the charge on the proton is the same as that on the electron (but opposite in sign), a neutral atom will contain the same number of protons in its nucleus as it has electrons in its orbit. The numbers of neutrons that go with the protons can vary by a little, giving the different isotopes of the atom.

However, the story does not stop at this point. Just as we once believed that the atom was fundamental and then discovered that it is composed of protons, neutrons and electrons, we now know that the protons and neutrons are not fundamental either (but the electron is, remember).

Protons and neutrons are composed of quarks.

Specifically, the proton is composed of two up quarks and one down quark. The neutron is composed of two down quarks and one up quark. Symbolically we can write this in the following way:

$$p \equiv uud$$
$$n \equiv udd.$$

As the proton carries an electrical charge, at least some of the quarks must also be charged. However, similar quarks exist inside the neutron, which is uncharged. Consequently the charges of the quarks must add up in the combination that composes the proton but cancel out in the combination that composes the neutron. Calling the charge on an up quark Q_u and the charge on a down quark Q_d, we have:

$$p \text{ (uud) charge} = Q_u + Q_u + Q_d = 1$$
$$n \text{ (udd) charge} = Q_u + Q_d + Q_d = 0.$$

Notice that in these relationships we are using a convention that sets the charge on the proton equal to $+1$. In standard units this charge would be approximately 1.6×10^{-19} coulombs. Particle physicists normally use this abbreviated unit and understand that they are working in multiples of the proton charge (the proton charge is often written as $+e$).

These two equations are simple to solve, producing:

$$Q_u = \text{charge on the up quark} = +\tfrac{2}{3}$$
$$Q_d = \text{charge on the down quark} = -\tfrac{1}{3}.$$

Until the discovery of quarks, physicists thought that electrical charge could only be found in multiples of the proton charge. The standard model suggests that there are three basic quantities of charge: $+2/3$, $-1/3$, found on the quarks, and -1 found on the electron (and other charged leptons)[2].

The other quarks also have charges of $+2/3$ or $-1/3$. Table 1.2 shows the standard way in which the quarks are grouped into families. All the quarks in the top row have charge $+2/3$, and all those in the bottom row have charge $-1/3$. Each column is referred to as a *generation*. The up and down quarks are in the first generation, the top and bottom quarks belong to the third generation.

Table 1.2 The grouping of quarks into generations (NB: the letters in brackets are the standard abbreviations for the names of the quarks).

	1st generation	2nd generation	3rd generation
$+2/3$	up (u)	charm (c)	top (t)
$-1/3$	down (d)	strange (s)	bottom (b)

This grouping of quarks into generations roughly follows the order in which they were discovered, but it has more to do with the way in which the quarks respond to the fundamental forces.

All the matter that we see in the universe is composed of atoms—hence protons and neutrons. Therefore the most commonly found quarks in the universe are the up and down quarks. The others are rather more massive (the mass of the quarks increases as you move from generation 1 to generation 2 and to generation 3) and very much rarer. The other four quarks were discovered by physicists conducting experiments in which particles were made to collide at very high velocities, producing enough energy to make the heavier quarks.

In the modern universe heavy quarks are quite scarce outside the laboratory. However, earlier in the evolution of the universe matter was far more energetic and so these heavier quarks were much more common and had significant roles to play in the reactions that took place. This is one of the reasons why particle physicists say that their experiments allow them to look back into the history of the universe.

We should now consider the leptons. One of the leptons is a familiar object—the electron. This helps in our study of leptons, as the properties of the electron are mirrored in the muon and the tau. Indeed, when the muon was first discovered a famous particle physicist was heard to remark 'who ordered that?'. There is very little, besides mass, that distinguishes the electron from the muon and the tau. They all have the same electrical charge and respond to the fundamental forces in the same way. The only obvious difference is that the muon and the tau are allowed to decay into other particles. The electron is a stable object.

Aside from making the number of leptons equal to the number of quarks, there seems to be no reason why the heavier leptons should exist. It is a matter of satisfaction to physicists that there are equal numbers of quarks and leptons, but there is no clear idea at this stage why this should be so. This 'coincidence' has suggested many areas of research that are being explored today.

The other three leptons are all called neutrinos as they are electrically neutral. This is not the same as saying, for example, that the neutron has a zero charge. A neutron is made up of three quarks. Each of these quarks carries an electrical charge. When a neutron is observed from a distance, the electromagnetic effects of the quark charges balance out making the neutron look like a neutral object. Experiments that probe inside the neutron can resolve the presence of charged objects within it. Neutrinos, on the other hand, are fundamental particles. They have no components inside them—they are *genuinely* neutral. To distinguish such particles from ones whose component charges cancel, we shall say that the neutrinos (and particles like them) are *neutral*, and that neutrons (and particles like them) have *zero charge*.

Neutrinos have extremely small masses, even on the atomic scale. Experiments with the electron-neutrino suggest that its mass is less than one ten-thousandth of that of the electron. Many particle physicists believe that the neutrinos have no mass at all. This makes them the most ghost-like objects in the universe. Many people are struck by the fact that neutrinos have no charge or mass. This seems to deny them any physical existence at all! However, neutrinos do have energy and this energy gives them reality.

The names chosen for the three neutrinos suggest that they are linked in some way to the charged leptons. The link is formed by the ways in which the leptons respond to one of the fundamental forces. This allows us to group the leptons into generations as we did with the quarks. Table 1.3 shows the lepton generations.

Table 1.3 The grouping of leptons into generations (NB: the symbols in brackets are the standard abbreviations for the names of the leptons).

	1st generation	2nd generation	3rd generation
-1	electron (e^-)	muon (μ^-)	tau (τ^-)
0	electron-neutrino (ν_e)	muon-neutrino (ν_μ)	tau-neutrino (ν_τ)

The masses of the leptons increase as we move up the generations (at least this is true of the top row; as noted above, it is still an open question whether the neutrinos have any mass at all).

At this stage we need to consider the forces by which all these fundamental particles interact. This will help to explain some of the reasons for grouping them in the generations (which, incidentally, will make the groups much easier to remember).

1.2   The four fundamental forces

A fundamental force cannot be explained as arising from the action of a more basic type of force. There are many forces in physics that occur in different situations. For example:

- gravity;
- friction;
- tension;
- electromagnetic[3];
- van der Waals.

Only two of the forces mentioned in this list (gravity and electromagnetic) are regarded as fundamental forces. The rest arise due to more fundamental forces.

For example, friction takes place when one object tries to slide over the surface of another. The theory of how frictional forces arise is very complex, but in essence they are due to the electromagnetic forces between the atoms of one object and those of another. Without electromagnetism there would be no friction.

Similarly, the tensional forces that arise in stretched wires are due to electromagnetic attractions between atoms in the structure of the wire. Without electromagnetism there would be no tension. Van der Waals forces are the complex forces that exist between atoms or molecules. It is the van der Waals attraction between atoms or molecules in a gas that allow the gas to be liquefied under the right conditions of temperature and pressure. These forces arise as a combination of the electromagnetic repulsion between the electrons of one atom and the electrons of another and the attraction between the electrons of one atom and the nucleus of another. Again the theory is quite complex, but the forces arise out of the electromagnetic force in a complex situation. Without the electromagnetic force there would be no van der Waals forces.

These examples illustrate the difference between a force and a fundamental force. Just as a fundamental particle is one that is not composed of any pieces, a fundamental force is one that does not arise out of a more basic force.

Particle physicists hope that one day they will be able to explain all forces out of the action of just one fundamental force. In chapter 11 we will see how far this aim has been achieved.

The standard model recognizes four forces as being sufficiently distinct and basic to be called fundamental forces:

- gravity;
- electromagnetic;
- the weak force;
- the strong force.

Our experiments indicate that these forces act in very different ways from each other at the energies that we are currently able to achieve. However, there is some theoretical evidence that in the early history of the universe particle reactions took place at such high energies that the forces started to act in very similar ways. Physicists regard this as an indication that there is one force, more fundamental than the four listed above, that will eventually be seen as the single force of nature.

Gravity and electromagnetism will already be familiar to you. The weak and strong forces may well be new. In the history of physics they have only recently been discovered. This is because these forces have a definite *range*—they only act over distances smaller than a set limit. For distances greater than this limit the forces become so small as to be undetectable.

The range of the strong force is 10^{-15} m and that of the weak force 10^{-17} m. A typical atom is about 10^{-10} m in diameter, so these forces have ranges smaller than atomic sizes. A proton on one side of a nucleus would be too far away from a proton on the other side to interact through the action of the weak force! Only in the last 60 years have we be able to conduct experiments over such short distances and observe the physics of these forces.

1.2.1 The strong force

Two protons placed 1 m apart from each other would electromagnetically repel with a force some 10^{42} times greater than the gravitational attraction between them. Over a similar distance the strong force would be zero. If, however, the distance were reduced to a typical nuclear diameter, then the strong force would be at least as big as the electromagnetic. It is the strong force attraction that enables a nucleus, which packs protons into a small volume, to resist being blown apart by electrostatic repulsion.

The strong force only acts between quarks. The leptons do not experience the strong force at all. They are effectively blind to it (similarly a neutral object does not experience the electromagnetic force). This is the reason for the division of the material particles into the quarks and leptons.

> Quarks feel the strong force, leptons do not.
> Both quarks and leptons feel the other three forces.

This incredibly strong force acting between the quarks holds them together to form objects (particles) such as the proton and the neutron. If the leptons could feel the strong force, then they would also bind together into particles. This is the major difference between the properties of the quarks and leptons.

> The leptons do not bind together to form particles.
> The strong force between quarks means that they can
> *only* bind together to form particles.

Current theories of the strong force suggest that it is impossible to have a single quark isolated without any other quarks. All the quarks in the universe at the moment are bound up with others into particles. When we create new quarks in our experiments, they rapidly combine with others. This happens so quickly that it is impossible to ever see one on its own.

The impossibility of finding a quark on its own makes them very difficult objects to study. Some of the important experimental techniques used are discussed in chapter 10.

1.2.2 The weak force

The weak force is the most difficult of the fundamental forces to describe. This is because it is the one that least fits into our typical imagination of what a force should do. It is possible to imagine the strong force as being an attractive force between quarks, but the categories 'attractive' and 'repulsive' do not really fit the weak force. This is because it changes particles from one type to another.

The weak force is the reason for the generation structure of the quarks and leptons. The weak force is felt by both quarks and leptons. In this respect it is the same as the electromagnetic and gravitational forces— the strong force is the only one of the fundamental forces that is only felt by one class of material particle.

If two leptons come within range of the weak force, then it is possible for them to be changed into other leptons, as illustrated in figure 1.1.

Figure 1.1 is deliberately suggestive of the way in which the weak force operates. At each of the black blobs a particle has been changed from one type into another. In the general theory of forces (discussed in chapter 11) the 'blobs' are called *vertices*. The weak force can change

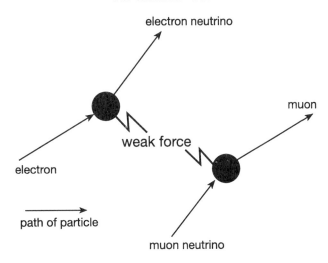

Figure 1.1 A representation of the action of the weak force.

a particle from one type into another at such a vertex. However, it is only possible for the weak force to change leptons *within the same generation* into each other. The electron can be turned into an electron-neutrino, and vice versa, but the electron cannot be turned into the muon-neutrino (or a muon for that matter). This is why we divide the leptons into generations. The weak force can act within the lepton generations, but not between them.

There is a slight complication when it comes to the quarks. Again the weak force can turn one quark into another and again the force acts within the generations of quarks. However, it is not true to say that the force cannot act across generations. It can, but with a much reduced effect. Figure 1.2 illustrates this.

The generations are not as strictly divided in the case of quarks as in the case of leptons—there is less of a generation gap between quarks.

This concludes a very brief summary of the main features of the four fundamental forces. They will be one of the key elements in our story and we will return to them in increasing detail as we progress.

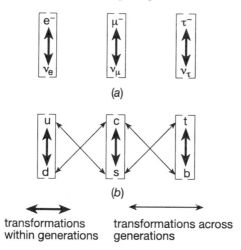

(a)

(b)

⟷ transformations within generations

⟷ transformations across generations

Figure 1.2 (a) The effect of the weak force on the leptons. (b) The effect of the weak force on the quarks (NB: transformations such as u → b are also possible).

1.3 The big bang

Our planet is a member of a group of nine planets that are in orbit round a star that we refer to as the sun. This family of star and planets we call our solar system. The sun belongs to a galaxy of stars (it sits on the edge of the galaxy, out in the suburbs) many of which may have solar systems of their own. In total there are something like 10^{11} stars in our galaxy. We call this galaxy the *Milky Way*.

The Milky Way is just one of many galaxies. It is part of the 'local group', a collection of galaxies held together by gravity. There are 18 galaxies in the local group. Astronomers have identified other clusters of galaxies, some of which contain as many as 800 separate galaxies loosely held together by gravity.

It is estimated that there are 3×10^9 galaxies observable from earth. Many of these are much bigger than our Milky Way.

The whole collection forms what astronomers call the observable universe—they refer to it as the observable universe out of caution:

there may be other bits that we have not seen yet! I shall just use the term 'universe' in this book—it is shorter. Anyway, we will have enough trouble with the bits that we can see without worrying about other unobserved parts as well.

The number of galaxies and the number of stars in the universe takes us beyond the point at which numbers have a meaning that can be grasped by the imagination. There are more stars in the universe than there are grains of sand on a beach. The physical size of the universe is also beyond imagination. A common distance unit used by astronomers is the light year. Some people are confused and think that a light year is a measurement of time, it is not—light years measure *distance*. One light year is the distance travelled by a beam of light in one year. As the speed of light is three hundred million metres per second, a simple calculation tells us that a light year is a distance of 9.45×10^{15} m. On this scale the universe is thought to be roughly 2×10^{10} light years in diameter.

Cosmology is the branch of physics that studies the possible ways in which the universe could have come about. It is not a normal branch of physics, in the sense that experimental confirmation of theory is not easily arranged![4] This does not mean that it is less successful or less rigorous, just harder. With contributions from the theories of astronomy and particle physics, cosmologists are now confident that they have a 'standard model' of how the universe began. This model is called the *big bang*.

Imagine a time some fifteen billion years ago. All the matter in the universe exists in a volume very much smaller than it does now— smaller, even, than the volume of the earth. There are no galaxies, indeed no matter as we would recognize it at all.

The temperature of the whole universe at this time is incredible, greater than 10^{13} K. The only material objects are elementary particles reacting with each other more often and at higher energies than we have ever been able to reproduce in our experiments (at such temperatures the kinetic energy of a single particle is greater than that of a jet plane).

And the universe is expanding. It is growing in size as all the particles fly away from each other at immense speeds. The universe cools as it expands. Each particle loses energy as the gravity of the rest of the universe pulls it back. Eventually the matter cools enough for the particles to combine in ways that we shall discuss as this book progresses—matter as we know it is formed. This matter cools some more and stars and galaxies start to form. Eventually man evolves on his little semi-detached planet.

All this time as matter is combining in small groups, the groups are still moving away from each other. The universe continues to grow, although more slowly now. We can see evidence for this expansion in the motion of the galaxies.

Why is the universe growing? Cosmologists believe that if we were to look far enough back in time we would see that the universe started as a point of immensely hot, unbelievably compressed matter that 'exploded'—the big bang. Evidence for this 'explosion' can be seen in the galaxies that are still flying apart today.

What caused the big bang to happen? We do not know. Our theories of how matter should behave do not work at the temperatures and pressures that existed just after the big bang. At such small volumes all the particles in the universe were so close to each other that gravity plays a major role in how they would react. As yet we do not know how to put gravity into our theories of particle physics. We do know that Newton's theory of gravity does not work in these circumstances; neither does that of Einstein.

There is a point in the history of the universe earlier than which our current best theories break down. This point is about 10^{-43} seconds after the big bang. The remarkable thing is that this number is so small!

Cosmologists tend to get tied up in the fact that we do not understand what happened before this. They are not often heard expressing amazement at the fact that we can get so close to the instant of creation (the point in time at which everything (time included!) started).

Particle physics is the key to this. At this early stage of the universe's existence, the elementary particles were the only recognizable forms of matter in the universe and they were reacting according to the rules and laws that we have discovered in our laboratories. These rules had a distinct impact on the way in which the universe developed. Their influence can be seen in the structure of the universe that we see today.

1.4   Summary of chapter 1

- There are two types of fundamental material particle: quarks and leptons;
- fundamental particles do not contain any other objects within them;
- there are four fundamental forces: strong, weak, electromagnetic and gravity;
- fundamental forces are not the result of simpler forces acting in complicated circumstances;
- there are six different quarks and six different leptons;
- the quarks can be divided into three pairings called generations;
- the leptons can also be divided into three generations;
- the quarks feel the strong force, the leptons do not;
- both quarks and leptons feel the other three forces;
- the strong force binds quarks together into particles;
- the weak force can turn one fundamental particle into another: in the case of leptons it can act only within generations, in the case of the quarks it predominantly acts within generations but can also act between them;
- the weak force cannot turn quarks into leptons or vice versa;
- the universe was, we believe, created some 15 billion years ago;
- the event of creation was a gigantic 'explosion'—the big bang—in which all the elementary particles were produced;
- as a result of this 'explosion' the bulk of the matter in the universe is flying apart, even today;
- the laws of particle physics determine the early history of the universe.

Notes

[1] Physicists are not very good at naming things. Over the past few decades there seems to have been an informal competition to see who can come up with the silliest name for a property or a particle. This is all harmless fun. However, it does create the illusion that physicists do not take their jobs seriously. Being

semi-actively involved myself, I am delighted that physicists are able to express their humour and pleasure in the subject in this way—it has been stuffy for far too long! However, I do see that it can cause problems for others. Just remember that the actual names are not important—it is what the particles *do* that counts!

Murray Gell-Mann has been at the forefront of the 'odd names' movement for several years. He has argued that during the period when physicists tried to name things in a meaningful way they invariably got it wrong. For example atoms, so named because they were indivisible, were eventually split. His use of the name 'quark' was a deliberate attempt to produce a name that did not mean anything, and so could not possibly be wrong in the future! The word is actually taken from a quotation from James Joyce's *Finnegan's Wake*: 'Three quarks for Muster Mark'.

[2] It is a very striking fact that the total charge of 2u quarks and 1d quark (the proton) should be exactly the same size as the charge on the electron. This is very suggestive of some link between the quarks and leptons. There are some theories that make a point of this link, but as yet there is no experimental evidence to support them.

[3] The electromagnetic force is the name for the combined forces of electrostatics and magnetism. The complete theory of this force was developed by Maxwell in 1864. Maxwell's theory drew all the separate areas of electrostatics, magnetism and electromagnetic induction together, so we now tend to use the terms electromagnetism or electromagnetic force to refer to all of these effects.

[4] I have seen a spoof practical examination paper that included the question: 'given a large energy source and a substantial volume of free space, prepare a system in which life will evolve within 15 billion years'.

Chapter 2

Aspects of the theory of relativity

In this chapter we shall develop the ideas of special relativity that are of most importance to particle physics. The aim will be to understand these ideas and their implications, not to produce a technical derivation of the results. We will not be following the historically correct route. Instead we will consider two experiments that took place after Einstein published his theory.

Unfortunately there is not enough space in a book specifically about particle physics to dwell on the strange and profound features of relativity. Instead we shall have to concentrate on those aspects of the theory that are specifically relevant to us. This is a more mathematical chapter than any of the others in the book. Those readers whose taste does not run to algebraic calculation can simply miss out the calculations in boxes, at least on first reading.

2.1 Momentum

The first experiment that we are going to consider was designed to investigate how the momentum of an electron varied with its velocity. We will use the results to explain one of the cornerstones of relativity.

Momentum as defined in Newtonian mechanics is the product of an object's mass and its velocity:

$$\text{momentum} = \text{mass} \times \text{velocity}$$
$$p = mv. \tag{2.1}$$

In a school physics experiment, momentum would be obtained by measuring the mass of an object when stationary, measuring its velocity while moving and multiplying the two results together. Experiments carried out in this way invariably demonstrate Newton's second law of motion in the form:

$$\text{applied force} = \text{rate of change of momentum}$$
$$f = \frac{\Delta(mv)}{\Delta t}. \tag{2.2}$$

However, there is a way to measure the momentum of a charged particle *directly*, rather than via measuring its mass and velocity separately. The technique relies on the force exerted on a charged particle moving through a magnetic field.

A moving charged particle passing through a magnetic field experiences a force determined by the charge of the particle, the speed with which it is moving and the size of the magnetic field. Specifically:

$$F = Bqv \tag{2.3}$$

where B = size of magnetic field, q = charge of particle, v = speed of particle. The direction of this force is given by Fleming's left-hand rule (figure 2.1). The rule is defined for a positively charged particle. If you are considering a negative particle, then the direction of motion must be reversed—i.e. a negative particle moving to the right is equivalent to a positive particle moving to the left.

If the charged particle enters the magnetic field travelling at right angles to the magnetic field lines, the force will always be at right angles to both the field and the direction of motion (figure 2.2). Any force that is always at $90°$ to the direction of motion is bound to move an object in a circular path. The charged particle will be deflected as it passes through the magnetic field and will travel along the arc of a circle.

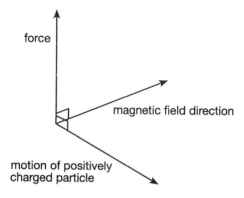

Figure 2.1 The left-hand rule showing the direction of force acting on a moving charge.

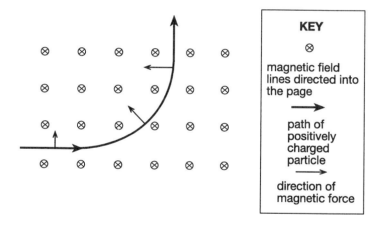

Figure 2.2 The motion of a charged particle in a magnetic field.

The radius of this circular path is a direct measure of the momentum of the particle. This can be shown in the context of Newtonian mechanics to be:

$$r = \frac{p}{Bq}$$

where r = radius of the path, p = momentum of the particle, B = magnetic field strength. A proof of this formula follows for those of a mathematical inclination.

To move an object of mass m on a circular path of radius r at a speed v, a force must be provided of size:

$$F = \frac{mv^2}{r}.$$

In this case, the force is the magnetic force exerted on the charged particle

$$F = Bqv$$

therefore

$$Bqv = \frac{mv^2}{r}$$

making r the subject:

$$r = \frac{mv}{Bq}$$

or

$$r = \frac{p}{Bq}$$

where p is the momentum of the particle. Equally:

$$p = Bqr.$$

This result shows that measuring the radius of the curve on which the particle travels, r, will provide a direct measure of the momentum, p, provided the size of the magnetic field and the charge of the particle are known.

This technique is used in particle physics experiments to measure the momentum of particles. Modern technology allows computers to reproduce the paths followed by particles in magnetic fields. It is then a simple matter for the software to calculate the momentum. When such experiments were first done (1909), much cruder techniques had to be used. The tracks left by the particles as they passed through photographic films were measured by hand to find the radius of the tracks. A student's thesis could consist of the analysis of a few such photographs.

The basis of the experiment is therefore quite simple: accelerate electrons to a known speed and let them pass through a magnetic field to measure their momentum. Electrons were used as they are lightweight particles with an accurately measured electrical charge.

Graph 2.1 shows the data produced by a series of experiments carried out between 1909 and 1915[1]. The results are startling.

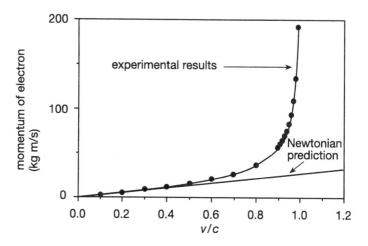

Graph 2.1 Momentum of an electron as a function of speed (NB: the horizontal scale is in fractions of the speed of light, c).

The simple Newtonian prediction shown on the graph is quite evidently wrong. For small velocities the experiment agrees well with the Newtonian formula for momentum, but as the velocity starts to get larger the difference becomes dramatic. The most obvious disagreement lies close to the magic number 3×10^8 m s^{-1}, the speed of light.

There are two possibilities here:

(1) the relationship between the momentum of a charged particle and the radius of curvature breaks down at speeds close to that of light;
(2) the Newtonian formula for momentum is wrong.

Any professional physicist would initially suspect the first possibility above the second. Although this is a reasonable suspicion, it is wrong. It is the second possibility that is true.

The formula that correctly follows the data is:

$$p = \frac{mv}{\sqrt{1 - v^2/c^2}} \qquad (2.4)$$

where p is the relativistic momentum. In this equation c^2 is the velocity of light squared, or approximately 9.0×10^{16} m² s⁻², a huge number! Although this is not a particularly elegant formula it does follow the data very accurately. This is the momentum formula proposed by the theory of relativity.

This is a radical discovery. Newtonian momentum is *defined* by the formula mv, so it is difficult to see how it can be *wrong*. Physicists adopted the formula $p = mv$ because it seemed to be a good description of nature; it helped us to calculate what might happen in certain circumstances.

One of the primary reasons why we consider momentum to be useful is because it is a *conserved quantity*. Such quantities are very important in physics as they allow us to compare situations before and after an interaction, without having to deal directly with the details of the interaction.

Consider a very simple case. Figure 2.3 shows a moving particle colliding and sticking to another particle that was initially at rest.

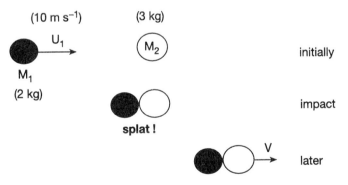

Figure 2.3 Collision between two particles that stick together.

The way in which particles stick together can be very complex and to study it in depth would require a sophisticated understanding of the molecular structure of the materials involved. However, if we simply want to calculate the speed at which the combined particles are moving after they stick together, then we can use the fact that momentum is conserved. The total momentum before the collision must be the same as that after the collision.

Using the Newtonian momentum formula, we can calculate the final speed of the combined particle in the following way:

$$\text{initial momentum} = M_1 U_1$$
$$= 2 \text{ kg} \times 10 \text{ m s}^{-1}$$
$$= 20 \text{ kg m s}^{-1}$$
$$\text{final momentum} = (M_1 + M_2)V$$
$$= 5 \text{ kg} \times V.$$

Momentum is conserved, therefore

$$\text{initial momentum} = \text{final momentum}.$$

Therefore

$$5 \text{ kg} \times V = 20 \text{ kg m s}^{-1}$$
$$V = \frac{20 \text{ kg m s}^{-1}}{5 \text{ kg}}$$
$$= 4 \text{ m s}^{-1}.$$

It is easy to imagine doing an experiment that would confirm the results of this calculation. On the basis of such experiments, we accept as a law of nature that momentum *as defined by* $p = mv$ is a conserved quantity. If mv were not conserved, then we would not bother with it—it would not be a useful number to calculate.

However, modern experiments show that it does not give the correct answers if the particles are moving quickly. We have seen how the magnetic bending starts to go wrong, but this can also be confirmed by simple collision experiments like that shown in figure 2.3. They also go wrong when the speeds become significant fractions of the speed of light.

For example, if the initially moving particle has a velocity of half the speed of light ($c/2$), then after it has struck the stationary particle the combined object should have a speed of $c/5$ ($0.2c$). However, if we carry out this experiment, then we find it to be $0.217c$. The Newtonian formula is *wrong*—it predicted the incorrect velocity. The correct velocity can be obtained by using the relativistic momentum instead of the Newtonian one.

Experiments are telling us that the quantity mv does not always correspond to anything useful—it is not always conserved. When the velocity of a particle is small compared with light (and the speed of light is so huge, so that covers most of our experience!), the Newtonian formula *is a good approximation to the correct one*[2]. This is why we continue to use it and to teach it. The correct formula is rather tiresome to use, so when the approximation $p \approx mv$ gives a good enough result it seems silly not to use it. The mistake is to act as if it will *always* give the correct answer.

There are other quantities that are conserved (energy for one) in particle physics reactions. Before the standard model was developed much of particle physics theory was centred round the study of conserved quantities.

2.2 Kinetic energy

Our second experiment considers the kinetic energy of a particle. Conventional theory tells us that when an object is moved through a certain distance by the application of a force, energy is transferred to the object. This is known as 'doing work':

work done = force applied × distance moved.

i.e.

$$W = F \times X \tag{2.5}$$

The energy transferred in this manner increases the kinetic energy of the object:

work done = change in kinetic energy. (2.6)

In Newtonian physics it is a simple matter to calculate the change in the kinetic energy, T, from Newton's second law of motion and the momentum, p. The result of such a calculation is:

$$T = \tfrac{1}{2}mv^2 = \frac{p^2}{2m}. \tag{2.7}$$

We can check this formula by accelerating electrons in an electrical field.

In a typical experiment a sequence of, equally spaced, charged metal plates is used to generate an electrical field through which electrons are passed (figure 2.4). The electrical fields between the plates are uniform, so the force acting on the electrons between the plates is constant. The energy transferred is then force $\times$ distance, the distance being related to the number of pairs of plates through which the electrons pass.

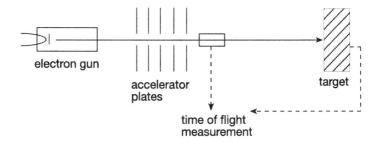

Figure 2.4 An energy transfer experiment.

The number of plates connected to the power supply can be varied, so the distance over which the electrons are accelerated can be adjusted. Having passed through the plates, the electrons coast through a fixed distance past two electronic timers that record their time of flight—hence their speed can be calculated. The whole of the interior of the experimental chamber is evacuated so the electrons do not lose any energy by colliding with atoms during their flight.

The prediction from Newtonian physics is quite clear. The work done on the electrons depends on the distance over which they are accelerated, the force being constant. If we assume that the electrons start from rest, then doubling the length over which they are accelerated by the electrical force will double the energy transferred, so doubling the kinetic energy. As kinetic energy depends on v^2, doubling the kinetic energy should also double v^2 for the electrons.

Graph 2.2 shows the variation of v^2 with the energy transferred to the electrons.

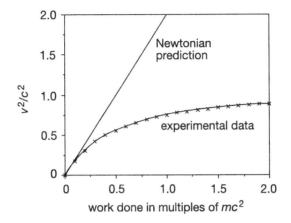

Graph 2.2 Kinetic energy related to work done.

Again the Newtonian result is very wrong. A check can be added to make sure that all the energy is actually being transferred to the electrons. The electrons can be allowed to strike a target at the end of their flight path. From the increase in the temperature of the target, the energy carried by the electrons can be measured. (In order for there to be a measurable temperature rise the experiment must run for some time so that many electrons strike the target.)

Such a modification shows that *all* the energy transferred is passing to the electrons, but the increase in the kinetic energy is not reflected in the expected increase in the velocity of the electrons. Kinetic energy does not depend on velocity in the manner predicted by Newtonian mechanics.

Einstein's theory predicts the correct relationship to be:

$$T = \frac{mc^2}{\sqrt{1 - v^2/c^2}} - mc^2 \qquad (2.8)$$

$$= (\gamma - 1)mc^2$$

using the standard abbreviation

$$\gamma = 1/\sqrt{1 - v^2/c^2}.$$

The Newtonian kinetic energy, T, is a good approximation to the true relativistic kinetic energy, $\boldsymbol{T}$, at velocities much smaller than light. We have been misled about kinetic energy in the same way as we were about momentum. Our *approximately* correct formula has led us into believing that it was the *absolutely* correct one. When particles start to move at velocities close to that of light, we must calculate their kinetic energies by using equation (2.8).

2.3   Energy

There is a curiosity associated with equation (2.8) that may have escaped your notice. One of the first things that one learns about energy in elementary physics is that absolute amounts of energy are unimportant, it is only *changes* in energy that matter[3]. If we consider equation (2.8) in more detail, then we see that it is composed of two terms:

$$\boldsymbol{T} = \gamma mc^2 - mc^2 \tag{2.8}$$

The first term contains all the variation with velocity (remember that γ is a function of velocity) and the second term is a fixed quantity that is always subtracted. If we were to accelerate a particle so that its kinetic energy increased from $\boldsymbol{T}_1$ to $\boldsymbol{T}_2$, say, then the change in kinetic energy (KE) would be:

$$\boldsymbol{T}_1 - \boldsymbol{T}_2 = (\gamma_1 mc^2 - mc^2) - (\gamma_2 mc^2 - mc^2)$$
$$= \gamma_1 mc^2 - \gamma_2 mc^2.$$

The second constant term, mc^2, has no influence on the change in KE. If we were to define some quantity $\boldsymbol{E}$ by the relationship:

$$\boldsymbol{E} = \gamma mc^2 \tag{2.9}$$

then the change in $\boldsymbol{E}$ would be the same as the change in KE.

The two quantities $\boldsymbol{T}$ and $\boldsymbol{E}$, differ only by a fixed amount mc^2. $\boldsymbol{T}$ and $\boldsymbol{E}$ are related by:

$$\boldsymbol{E} = \boldsymbol{T} + mc^2 \tag{2.10}$$

an obvious, but useful, relationship.

We can see from equation (2.10) that E is not zero even if T is. E is referred to as the *relativistic energy* of a particle. The relativistic energy is composed of two parts, the kinetic energy, T, and another term mc^2 that is not zero even if the particle is at rest. I shall refer to this second term as the *intrinsic energy* of the particle[4]. This intrinsic energy is related to the mass of the particle and so cannot be altered without turning it into a different particle.

2.4   Energy and mass

If $E = T + mc^2$, then a stationary particle ($T = 0$) away from any other objects (so it has no potential energy either) still has intrinsic energy mc^2. This strongly suggests that the intrinsic energy of a particle is deeply related to its mass. Indeed one way of interpreting this relationship is to say that all forms of energy have mass (i.e. energy can be weighed!). A metal bar that is hot (and so has thermal energy) would be slightly more massive than the same bar when cold[5].

Perhaps a better way of looking at it would be to say that energy and mass are different aspects of the same thing—relativity has blurred the distinction between them.

However one looks at it, one cannot escape the necessity for an intrinsic energy. Relativity has told us that this energy must exist, but provides no clues to what it is! Fortunately particle physics suggests an answer.

Consider a proton. We know that protons consist of quarks, specifically a uud combination. These quarks are in constant motion inside the proton, hence they have some kinetic energy. In addition, there are forces at work between the quarks—principally the strong and electromagnetic forces—and where there are forces there must be some potential energy (PE). Perhaps the intrinsic energy of the proton is simply the KE and PE of the quarks inside it?

Unfortunately, this is not the complete answer. There are two reasons for this:

(1) Some particles, e.g. the electron, do not have other particles inside them, yet they have intrinsic energy. What is the nature of this energy?

(2) The quarks themselves have mass, hence they have intrinsic energy of their own. Presumably, this quark energy must also contribute to the proton's intrinsic energy—indeed we cannot get the total intrinsic energy of the proton without this contribution. But then we are forced to ask the nature of the intrinsic energy of the quarks! We have simply pushed the problem down one level.

We can identify the contributions to the intrinsic energy of a composite particle:

$$\text{intrinsic energy of proton} = \text{KE of quarks} + \text{PE of quarks}$$
$$+ \text{intrinsic energy of quarks} \quad (2.11)$$

but we are no nearer understanding what this energy is if the particle has no internal pieces. Unfortunately, the current state of research has no conclusive answer to this question.

Of course, there are theories that have been suggested. The most popular is the so-called 'Higgs mechanism'. This theory is discussed in some detail later (page 205). However, there is little direct experimental evidence for this at present. If Higgs is correct, then a type of particle known as the Higgs boson should exist. So far it has not been found in experiments. There is a very active search going on at the moment. It is hoped that the recent upgrade to the LEP accelerator might provide enough energy to produce a Higgs particle, but we will probably have to wait for the Large Hadron Collider (LHC) which should begin operation in 2005[6].

2.4.1 Photons

The nature of the photon is an excellent illustration of the importance of the relativistic energy, rather then the kinetic energy. Photons are bursts of electromagnetic radiation. When an individual atom radiates light it does so by emitting photons. Some aspects of a photon's behaviour are very similar to that of particles, such as electrons. On the other hand, some aspects of a photon's behaviour are similar to that of waves. For example, we can say that a photon has a wavelength! We shall see in the next chapter that electrons have some very odd characteristics as well.

Photons do not have electrical charge, nor do they have any mass. In a sense they are pure kinetic energy. In order to understand how this can be, consider equation (2.12), which can be obtained, after some algebraic manipulation, from the relativistic mass and energy equations:

$$E^2 = p^2c^2 + m^2c^4. \tag{2.12}$$

From this equation one can see that a particle with no mass can still have relativistic energy. If $m = 0$, then:

$$E^2 = p^2c^2 \quad \text{or} \quad E = pc. \tag{2.13}$$

If a particle has no mass then it has no intrinsic energy. However, as the relativistic energy is not *just* the intrinsic energy, being massless does not prevent it from having *any* energy. Even more curiously, having no mass does not mean that it cannot have momentum. From equation (2.13) $p = E/c$. Our new expanded understanding of momentum shows that an object can have momentum even if it does not have mass—you can't get much further from Newtonian $p = mv$ than that!

The photon is an example of such a particle. It has no intrinsic energy, hence no mass, but it does have kinetic (and hence relativistic) energy and it does have momentum.

However, at first glance this seems to contradict the relativistic momentum equation, for if:

$$p = \frac{mv}{\sqrt{1 - v^2/c^2}} \tag{2.4}$$

then $m = 0$ implies that $p = 0$ as well. However there is no contradiction in one specific circumstance. If $v = c$ (i.e. the particle moves at the speed of light) then:

$$\begin{aligned} p &= \frac{mv}{\sqrt{1 - v^2/c^2}} \\ &= \frac{0 \times c}{\sqrt{1 - c^2/c^2}} = \frac{0}{0} \end{aligned}$$

an operation that is not defined in mathematics. In other words, equation (2.4) does not apply to a massless particle moving at the speed of light. This loophole is exploited by nature as photons, gluons[7] and possibly neutrinos are all massless particles and so must move at the speed of light.

In 1900 Max Planck suggested that a photon of wavelength λ had energy E given by:

$$E = \frac{hc}{\lambda} \qquad (2.14)$$

where $h = 6.63 \times 10^{-34}$ J s. We now see that this energy is the relativistic energy of the photon, so we can also say that:

$$p = \frac{h}{\lambda} \qquad (2.15)$$

which is the relativistic momentum of a photon and all massless particles.

2.4.2 Mass again

This section can be omitted on first reading.

I would like to make some final comments regarding energy and mass. The equation $E = mc^2$ is widely misunderstood. This is partly because there are many different ways of looking at the theory of relativity. Particle physicists follow the sort of understanding that I have outlined. The traditional alternative is to consider equation (2.4) to mean that the mass of a particle varies with velocity, i.e.:

$$p = Mv \quad \text{where} \quad M = \frac{m}{\sqrt{1 - v^2/c^2}}.$$

In this approach, m is termed the *rest mass* of the particle. This has the advantage of explaining *why* Newtonian momentum is wrong—Newton did not know that mass was not constant.

Particle physicists do not like this. They like to be able to identify a particle by a *unique* mass, not one that is changing with speed. They also point out that the mass of a particle is not easily measured while it is moving, *only the momentum and energy of a moving particle are important quantities.* They prefer to accept the relativistic equations for energy and momentum and only worry about the mass of a particle when it is at rest. The equations are the same which ever way you look at them.

However, confusion arises when you start talking about $E = mc^2$: is the 'm' the rest mass or the mass of the particle when it is moving?

Particle physicists use $E = mc^2$ to calculate the relativistic energy of a particle at rest—in other words the intrinsic energy. If the particle is moving, then the equation should become $E = T + mc^2 = \gamma mc^2$. In other words they reserve 'm' to mean *the* mass of the particle.

2.4.3 Units

The SI unit of energy is the *joule* (defined as the energy transferred when a force of 1 newton acts over a distance of 1 metre). However, in particle physics the joule is too large a unit to be conveniently used (it is like measuring the length of your toenails in kilometres—possible, but far from sensible), so the *electron-volt* (eV) is used instead. The electron-volt is defined as the energy transferred to a particle with the same charge as an electron when it is accelerated through a potential difference (voltage) of one volt.

The formula for energy transferred by accelerating charges through voltages is:

$$E = QV$$

where E = energy transferred (in joules), Q = charge of particle (in coulombs), V = voltage (in volts), so for the case of an electron and 1 V the energy is:

$$\begin{aligned}
E &= 1.6 \times 10^{-19} \text{ C} \times 1 \text{ V} \\
&= 1.6 \times 10^{-19} \text{ J} \\
&= 1 \text{ eV}.
\end{aligned}$$

This gives us a conversion between joules and electron-volts. In particle physics the eV turns out to be slightly too small, so we settle for GeV (G being the abbreviation for giga, i.e. 10^9).

Particle physicists have also decided to make things slightly easier for themselves by converting all their units to multiples of the velocity of light, c. In the equation:

$$E^2 = p^2 c^2 + m^2 c^4 \qquad (2.12)$$

every term must be in the same units, that of energy (GeV). Hence to make the units balance momentum, p, is measured in GeV/c and mass is measured in GeV/c^2. This seems quite straightforward until you realize *that we do not bother to divide by the actual numerical value of c*. This can be quite confusing until you get used to it. For example a proton at rest has intrinsic energy = 0.938 GeV and mass 0.938 GeV/c^2 *the same numerical value in each case.* You have to read the units carefully to see what is being talked about!

The situation is sometimes made worse by the value of c being set as 1 and all the other units altered to come into line. Energy, mass and momentum then all have the same unit, and equation (2.12) becomes:

$$E^2 = p^2 + m^2. \qquad (2.16)$$

This is a great convenience for people who are used to working in 'natural units', but can be confusing for the beginner. In this book we will use equation (2.12) rather than the complexities of 'standard units'.

2.5 Reactions and decays

2.5.1 Particle reactions

Many reactions in particle physics are carried out in order to create new particles. This is taking advantage of the link between intrinsic energy and mass. If we can 'divert' some kinetic energy from particles into intrinsic energy, then we can create new particles. Consider this reaction:

$$p + p \rightarrow p + p + \pi^0. \qquad (2.17)$$

This is an example of a *reaction equation*. On the left-hand side of the equation are the symbols for the particles that enter the reaction, in this case two protons, p. The '+' sign on the left-hand side indicates that the protons came within range of each other and the strong force triggered a reaction. The result of the reaction is shown on the right-hand side. The protons are still there, but now a new particle has been created as well—a neutral pion, or pi zero, π^0.

The arrow in the equation represents the boundary between the 'before' and 'after' situations.

This equation also illustrates the curious convention used to represent a particle's charge. The '0' in the pion's symbol shows that it has a zero charge. The proton has charge of $+1$ so it should be written 'p$^+$'. However, this has never caught on and the proton is usually written as just p. Most other positively charged particles carry the '+' symbol (for example the positive pion or π^+). Some neutral particles carry a '0' on their symbols and some do not (for example the Ξ^0 and the n). Generally speaking, the charge is left off when there is no chance of confusion—there is only one particle whose symbol is p, but there are three different π particles so in the latter case the charges must be included.

All reactions are written as reaction equations following the same basic convention:

$$A + B \rightarrow C + D$$

$$A \text{ hits } B \overset{(\text{becoming})}{\rightarrow} C \text{ and } D.$$

Reaction (2.17) can only take place if there is sufficient *excess* energy to create the intrinsic energy of the π^0 according to $E = mc^2$. I have used the term 'excess energy' to emphasize that only part of the energy of the protons can be converted into the newly created pion.

One way of carrying out this reaction is to have a collection of stationary protons (liquid hydrogen will do) and aim a beam of accelerated protons into them. This is called a fixed target experiment. The incoming protons have momentum so even though the target protons are stationary, the reacting combination of particles contains a net amount of momentum. Hence the particles leaving the reaction must be moving in order to carry between them the same amount of momentum, or the law of conservation would be violated. If the particles are moving then there must be some kinetic energy present after the reaction. This kinetic energy can only come from the energy of the particles that entered the reaction. Hence, not all of the incoming energy can be turned into intrinsic energy of new particles, the rest must be kinetic energy spread among the particles:

$$2 \times (\text{protons KE} + \text{IE}) = 2 \times (\text{protons KE} + \text{IE}) + (\pi^0\text{'s KE} + \text{IE})$$

where KE = kinetic energy and IE = internal energy. If the reacting particles are fired at each other with the same speed then the net momentum is zero initially. Hence, all the particles after the reaction could be stationary. This means that all the initial energy can be used to materialize new particles. Such 'collider' experiments are more energy efficient than fixed target experiments. They do have disadvantages as well, which we shall discuss in chapter 10.

In what follows I shall assume that the energy of the incoming particles is always sufficient to create the new particles and to conserve momentum as well.

2.5.2   Particle decays

Many particles are unstable, meaning that they will decay into particles with less mass. We would write such an event as:

$$A \rightarrow B + C$$

A being the decaying particle, and B and C the decay products (or daughters).

In such decays energy and momentum are also conserved. From the point of view of the decaying particle it is at rest when it decays, so there is no momentum. The daughter particles must then be produced with equal and opposite momenta (if there are only two of them). For example the π^0 decays into two photons:

$$\pi^0 \rightarrow \gamma + \gamma.$$

We can use the various relativistic equations to calculate the energy of the photons that are emitted.

From the decaying particle's point of view it is stationary, so only its intrinsic energy will be present initially. After the decay, the total energy of the two photons must be equal to the initial energy, i.e.:

$$\text{intrinsic energy of } \pi^0 = \text{total energy of photons}$$
$$M_{\pi^0}c^2 = 2E$$
$$\text{mass of } \pi^0 = 0.134 \text{ GeV}/c^2$$
$$\therefore \qquad 2E = (0.134 \text{ GeV}/c^2) \times c^2$$
$$E = 0.0567 \text{ GeV}.$$

The situation is slightly more complicated if the produced particles have mass as well. The best example of this situation is the decay of the K^0. It decays into two of the charged pions:

$$K^0 \rightarrow \pi^+ + \pi^-.$$

To calculate the energy of the pions we follow a similar argument. As the pions have the same mass as each other they are produced with the same energy and momentum. This makes the calculation quite simple.

$$\text{intrinsic energy of K} = \text{total energy of pions}$$
$$M_{K^0}c^2 = 2E_\pi$$
$$\therefore \qquad E_\pi = \frac{M_{K^0}c^2}{2}$$

as the mass of the $K^0 = 0.498 \text{ GeV}/c^2$

$$E_\pi = 0.249 \text{ GeV}.$$

We can go on to calculate the momenta of the produced pions:

$$E^2 = p^2c^2 + m^2c^4$$
$$(0.249 \text{ GeV})^2 = p^2c^2 + (0.140 \text{ GeV}/c^2)^2c^4$$
$$\therefore \qquad p^2c^2 = (0.249 \text{ GeV})^2 - (0.140 \text{ GeV}/c^2)^2c^4$$
$$= 0.042 \text{ (GeV)}^2$$
$$\therefore \qquad p = 0.206 \text{ GeV}/c.$$

2.6   Summary of chapter 2

- Momentum can be measured *directly* by magnetic bending;
- the formula $p = mv$ is only an approximation to the true relationship $p = \gamma mv$;
- relativistic momentum is the quantity that is really conserved in collisions;
- relativistic kinetic energy is given by $T = \gamma mc^2 - mc^2$;
- relativistic energy (a more useful quantity) is $E = \gamma mc^2$;
- mc^2 is the intrinsic energy of a particle which is responsible for giving the particle mass;
- particles that are massless must move at the speed of light;
- such particles (e.g. the photon and the neutrinos) can have energy and momentum even though they do not have mass;
- $E^2 = p^2c^2 + m^2c^4$;
- if $m = 0$, then $p = E/c$ and $E = pc$;
- when particles react excess energy can be diverted into intrinsic energy of new particles, subject to constraints imposed by conservation of energy and momentum;
- when particles decay energy and momentum are also conserved.

Notes

[1] The data are taken from the experiments of Kaufmann (1910), Bucherer (1909) and Guye and Lavanchy (1915).

[2] Consider what happens when the speed of the particle, v, is quite small compared with light. In that case the factor $v^2/c^2 \ll 1$ so that $(1 - v^2/c^2) \approx 1$ and $\sqrt{1 - v^2/c^2} \approx 1$, in which case the relativistic momentum $p \approx mv$!

[3] You may think that KE is an exception to this. After all, if an object is not moving then it has no KE, and there cannot be much dispute about that! But consider: any object that is not moving from one person's point of view, may well be moving from another person's viewpoint. Which is correct?

[4] There are various alternative names such as 'rest energy', but they are rooted in the idea of 'rest mass' which is a way of looking at relativity that is not generally used by particle physicists.

[5] The difference in mass is minute as c^2 is a very large number.

[6] In 1993, a proposed particle accelerator in America, the Superconducting Supercollider (SSC) was cancelled due to financial problems. The projected cost of the machine was $11 billion. One of the primary jobs of the SSC was to be the search for the Higgs boson.

[7] Gluons are the strong force equivalent of photons.

Chapter 3

Quantum theory

In this chapter we shall consider some aspects of quantum mechanics. The aim will be to outline the parts of quantum mechanics that are needed for an understanding of particle physics. We shall explore two experiments and their interpretation. The version of quantum mechanics that we shall develop (and there are many) will be that of Richard Feynman as this is the most appropriate for understanding the fundamental forces.

'...This growing confusion was resolved in 1925 or 1926 with the advent of the correct equations for quantum mechanics. Now we know how the electrons and light behave. But what can I call it? If I say they behave like particles I give the wrong impression; also if I say they behave like waves. They behave in their own inimitable way, which technically could be called a quantum mechanical way. They behave in a way that is like nothing that you have ever seen before. Your experience with things that you have seen before is incomplete. The behaviour of things on a very small scale is simply different. An atom does not behave like a weight hanging on a spring and oscillating. Nor does it behave like a miniature representation of the solar system with little planets going around in orbits. Nor does it appear to be like a cloud or fog of some sort surrounding the nucleus. It behaves like nothing you have ever seen before.

There is one simplification at least. Electrons behave in this respect in exactly the same way as photons; they are both screwy, but in exactly the same way...

The difficulty really is psychological and exists in the perpetual torment that results from your saying to yourself 'but how can it really be like that?' which is a reflection of an uncontrolled but vain desire to see it in terms of something familiar. I will not describe it in terms of an analogy with something familiar; I will simply describe it...

I am going to tell you what nature behaves like. If you will simply admit that maybe she does behave like this, you will find her a delightful and entrancing thing. Do not keep saying to yourself, if you can possibly avoid it, 'but how can it be like that?' because you will get 'down the drain', into a blind alley from which nobody has yet escaped. Nobody knows how it can be like that.'
From *The Character of Physical Law* by Richard P Feynman.

The 20th century has seen two great revolutions in the way we think about the physical world. The first, relativity, was largely the work of one man, Albert Einstein. The second, quantum theory, was the pooled efforts of a wide range of brilliant physicists.

The period between 1920 and 1935 was one of great turmoil in physics. A series of experiments demonstrated that the standard way of calculating, which had been highly successful since the time of Newton, did not produce the right answers when it was applied to atoms. It was as if the air that physicists breathe had been taken away. They had to learn how to live in a new atmosphere. Imagine playing a game according to the rules that you had been using for years, but suddenly your opponent started using different rules—and was not telling you what they were.

Gradually, thanks to the work of Niels Bohr, Werner Heisenberg, Max Born, Erwin Schrödinger and Paul Dirac (amongst others) the new picture started to emerge. The process was one of continual debate amongst groups of scientists, trying out ideas and arguing about them until a coherent way of dealing with the atomic world was produced.

Contrast this with the development of relativity: one man, working in a patent office in a burst of intellectual brilliance producing a series of papers that set much of the agenda for 20th century physics.

The history of the development of quantum mechanics is a fascinating human story in its own right. However, in this book we must content ourselves with developing enough of the ideas to help us understand the nature of the fundamental forces.

In the chapter on relativity I was able to concentrate on the results of two experiments that demonstrated some of the new features of the theory. In this chapter we shall also look at two experiments (out of historical context). However, much of the time we will be looking at how we interpret what the experiments mean to develop a feel for the bizarre world that quantum mechanics reveals to us.

3.1   The double slot experiment for electrons

A classic experiment that cuts to the heart of quantum mechanics is the double slot experiment for electrons. It is a remarkably simple experiment to describe (although somewhat harder to do in practice) considering how fundamentally it has shaken the roots of physics.

The idea is to direct a beam of electrons at a metal screen into which have been cut two rectangular slots. The slots are placed quite close to each other, closer than the width of the beam, and are very narrow (see figure 3.1). The electron beam can be produced in a very simple way using equipment similar to that found inside the tube of an ordinary television. However, an important addition is the ability to reduce the intensity of the beam so that the number of electrons per second striking the screen can be carefully controlled.

On the other side of the screen, and some distance away, is a device for detecting the arrival of electrons. This can be a sophisticated electronic device such as used in particle physics experiments, or it can be a simple photographic film (of very high sensitivity). The purpose of the experiment is to count the number of electrons arriving at different points after having passed through the slots in the screen.

The experiment is left running for some time and then the photographic film is developed. The exposed film shows a pattern of dark and light patches. In negative, the dark patches are where electrons have arrived and exposed the film. The darker the patch the greater the number

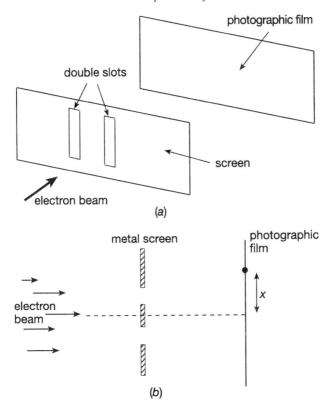

Figure 3.1   (a) The double slot experiment for electrons.   (b) The experiment seen from above.

of electrons that arrived there during the experiment. A series of such exposures can be used to estimate the probability of an electron arriving at any given point (such as that marked x on the diagram).

This is a very significant experiment as the results disagree totally with one's expectations. Common sense would have us argue along the following lines[1]:

- an electron will either strike the screen, in which case it will be absorbed, or pass through one of the slots (we can prevent the screen from charging up as electrons hit it by connecting it to ground);

- any electron that reaches the film on the other side of the screen must have passed through one or the other slot (the screen is so big compared to the beam that none can leak round the edge);
- if the film is 'uniform' i.e. every part of the film is equally sensitive to electrons arriving and no electrons are lost in transit (the experiment can be done in a vacuum to prevent electrons colliding with atoms in the air) then the film should record the arrival of every electron that gets through a slot;
- so if we could in some way 'label' the electrons as having come through slot 1 or slot 2, then the total number arriving at x will be the sum of those that arrived having come through slot 1 plus those that arrived after having passed through slot 2.

Now we cannot label electrons, but what we can do is to carry out the experiment with one or other of the slots blocked off and count the number of electrons arriving at x. In other words:

$$\begin{array}{c}\text{fraction of electrons arriving}\\\text{at } x \text{ with both slots open}\end{array} = \begin{array}{c}\text{fraction arriving}\\\text{with slot 1 open}\end{array} + \begin{array}{c}\text{fraction arriving}\\\text{with slot 2 open}\end{array}$$

The fraction of electrons arriving at a point is the experimental way in which the probability of arriving is measured. So, using the symbol $P_1(x)$ to mean the probability of an electron arriving at x on the screen after having passed through slot 1, etc, we can write:

$$P_{12}(x) = P_1(x) + P_2(x).$$

This is an experimental prediction that can be tested. It may strike you that it is such an obvious statement that there is no need to test it. Well, unfortunately it turns out to be wrong!

Figure 3.2 shows the distribution of electrons arriving at a range of points as measured with one or the other slots blocked off.

Both of these curves are exactly what one would expect after some thought. Most of the electrons that pass straight through arrive at the film at a point opposite the slot. They need not travel through the slot exactly parallel to the centre line (and some might ricochet off the edge of the slot) so the distribution has some width.

Now consider the experimental results obtained from carrying out the

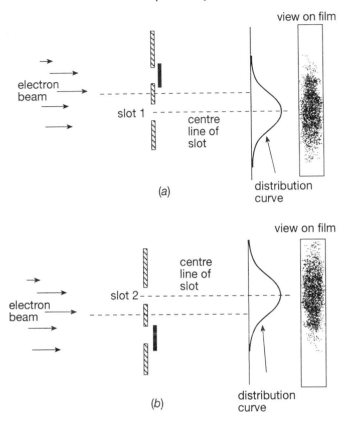

Figure 3.2 (a) The electron distribution for slot 1 open and slot 2 closed; the height of the curve indicates the fraction of electrons arriving at that point. (b) The distribution of electrons with slot 2 open and slot 1 closed.

experiment with both slots open at the same time. This is shown in figure 3.3.

No matter how one looks at this result, it is *not* the sum of the other two distributions. Consider in particular a point such as that labelled *y* on the diagram. Comparing the number of electrons arriving here when there are both slots open with the number when only slot 1 is open leads to the odd conclusion *that opening the second slot has reduced the number of electrons arriving at this point*! Indeed points on the film such as *y* show no darkening—electrons *never* arrive there[2].

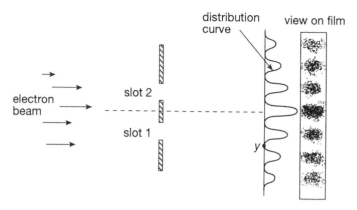

Figure 3.3 The distribution of electrons with both slots open.

As if that were not enough to cause a puzzle, consider the experiment from the point of view of an electron passing through slot 1. If this is the path followed by the electron, then how can it 'know' if slot 2 is open or closed? Yet it clearly can, in some sense. If slot 2 is open then electrons never arrive at a point such as y, yet if slot 2 is closed electrons frequently arrive at y.

As a final twist it is possible to reduce the intensity of the electron beam to such an extent that the time between electrons arriving at the screen is very much greater than the time it takes them to cross from the screen to the film. This ensures that only one electron at a time is passing through the equipment. We would want to do this to ensure that one electron passing through one slot does not have some effect on another passing through the other slot[3].

If we decide to develop the film after each electron arrives, then we should be able to trace the behaviour of a single electron (it is not possible to do this with film, but it can be done with electronic equipment). Unsurprisingly, every time we develop the film we find a single small spot corresponding to the arrival point of a single electron. However, if we stack the films on top of one another, then together all the single electrons draw out the distribution shown in figure 3.3.

You must let the significance of this result sink in. A single electron crossing the equipment will never arrive at one of the zero points in the distribution of figure 3.3. Yet common sense would dictate that a single electron must either go through slot 1 or slot 2 and so should arrive at some point with a relative probability given by either figure 3.2(a) or figure 3.2(b). Experimentally this has been shown to be incorrect. Every individual electron crossing the equipment 'knows' that both slots are open.

3.2 What does it all mean?

While the experimental facts outlined in the previous section have been established beyond doubt, their interpretation is very much an open question.

When physicists try to construct a theory from their experimental observations they are primarily concerned with two things:

- how to calculate something that will enable them to predict the results of further experiments
- how to understand what is going on in the experiment.

Now, it is not always possible to satisfy both of these concerns. Sometimes we have a reasonably clear idea what is going on, but the mathematical details are so complex that calculating is very hard. Sometimes the mathematics is quite clear, but the understanding is difficult.

Quantum mechanics falls into the latter category. As the pioneers in this field struggled with the results of experiments just as baffling as the double slot experiment, a very pragmatic attitude started to form. They began to realize that an understanding of *how* electrons could behave in such a manner would come once they developed the rules that would enable them to calculate the way they behaved. So quantum mechanics as a set of mathematical rules was developed. It has been tested in a huge number of experiments since then and never found to be wrong. Quantum mechanics as a theory works extremely well. Without it the microchip would not have been developed. Yet we are still not sure how electrons can behave in such a manner.

In some ways the situation has got worse over the years. Some people held out the hope that quantum mechanics was an incomplete theory and that as we did more experiments we would discover some loose end that allowed us to make sense of things. This has not happened. Indeed versions of the double slot experiment have been carried out with photons and, more recently, sodium atoms—with the same results.

If we accept that this is just the way things are and that, like it or not, the subatomic world does behave in such a manner then another question arises. Why do we not see larger objects, such as cricket balls, behaving in this manner? One could set up an experiment similar to the double slot (Feynman has suggested using machine gun bullets and armour plating!), yet one cannot believe that large objects would behave in such a manner. Indeed, they do not. To gain some understanding of why the size of the object should make such a difference, we need to develop some of the machinery of quantum mechanics.

3.3   Feynman's picture

In the late 1940s Richard Feynman decided that he could not relate to the traditional way of doing quantum mechanics and decided to rethink things for himself. He came up with a totally new way of doing quantum mechanical calculations coupled with a new picture of how to interpret the results of the experiments. His technique is highly mathematical and very difficult to apply to 'simple' situations such as electrons in atoms. However, when applied to particles and the fundamental forces the technique comes into its own and provides a beautifully elegant way of working. The Feynman method is now almost universally applied in particle physics, so it is his ideas that we shall be using rather than the earlier view developed by Heisenberg, Schrödinger and Bohr[4].

The starting point is to reject the relationship:

$$P_{12}(x) = P_1(x) + P_2(x).$$

After all, P_1 and P_2 have been obtained from two different experiments and, despite the urgings of common sense, there is no reason in principle why they should still apply in a different, third, experiment. Indeed, evidence suggests that they don't! Accordingly it is necessary to calculate, or measure, the probability for *each* experiment and to be very careful about carrying information from one to another.

In the first instance quantum mechanics does not work with probabilities, it works with mathematical quantities called *amplitudes*. Amplitudes are related to probabilities:

$$\text{probability of an event} = |\text{amplitude for an event}|^2.$$

but they are not the same thing. (The straight brackets, $|$, indicate that this is not a simple squaring as with numbers, e.g. $2^2 = 4$. This is known as the *absolute square*. The way in which the absolute square is calculated is described later.) Amplitudes cannot be measured directly. Only probabilities can be measured. However, it is the amplitudes that dictate the motion of particles.

The basic rules of quantum mechanics are quite clear about this. To calculate the probability of an event happening (an electron arriving at a point on a film) one must calculate the amplitude for it to happen, and then absolute square the result to get the probability. Why not just calculate the probability directly? Because amplitudes *add* differently to probabilities.

According to Feynman, to calculate the amplitude for an electron arriving at a point on a film one has to consider all the possible paths that the electron might have taken to get there. In figure 3.4 a couple of possible paths are drawn.

Feynman invented a method for calculating the amplitude for each possible path. To get the total amplitude one just adds up the contributions for each path. Writing the amplitude to get from a to b as $A(a, b)$:

$$A(a, b) = A(a, b)_{\text{path 1}} + A(a, b)_{\text{path 2}} + A(a, b)_{\text{path 3}} + \cdots$$

and

$$\text{probability}(a, b) = |A(a, b)_{\text{path 1}} + A(a, b)_{\text{path 2}} + A(a, b)_{\text{path 3}} + \cdots|^2.$$

Now, as it is the total amplitude that is absolute squared the result is not the same as taking the amplitudes separately. Consider the simple case of just two paths:

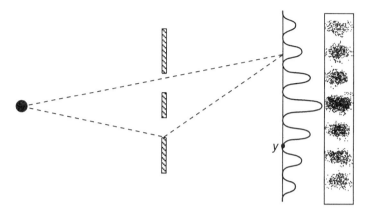

Figure 3.4 Two possible paths leading to the same point on the distribution.

- amplitude for path $1 = A_1$
- amplitude for path $2 = A_2$
- so, probability if only path 1 is available $= |A_1|^2$
- probability if only path 2 is available $= |A_2|^2$
- but, if both paths are available then:
  probability $= |A_1 + A_2|^2 = A_1^2 + A_2^2 + 2 \times A_1 \times A_2$
- and it is the extra piece in this case that makes the difference.

It does not matter how odd or unlikely the path seems to be, it will have an amplitude and its contribution must be counted. Sometimes this contribution will be negative. That will tend to reduce the overall amplitude. In some cases the amplitude is made up of contributions from different paths that totally cancel each other out (equal positive and negative parts). In this case the amplitude to get from that *a* to that *b* is zero—it never happens. Such instances are the zero points in the distribution of figure 3.3. Electrons never arrive at these points as the amplitudes for them to do so along different paths totally cancel each other out. Zero amplitude means zero probability.

Clearly, if one of the slots is blocked off then there are a whole range of paths that need not be considered in the total sum. This explains why the probability distributions look different when only one slot is open.

3.4   A second experiment

The second experiment that we need to consider involves scattering particles off one another (see figure 3.5). This will reveal some more interesting features of calculating with amplitudes.

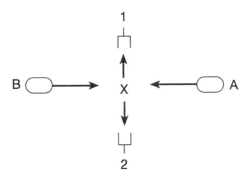

Figure 3.5 Scattering nuclei off each other.

The experimental equipment consists of two sources (placed at A and B) of two different nuclei, α and β. The sources are aimed at each other, and two detectors, 1 and 2, are placed at right angles to the path of the nuclei. The simple idea is to collide the nuclei at point X and count how many arrive at detector 1 and detector 2.

To calculate the probability that a nucleus from A arrives at 1 we need to calculate the amplitude for it to cross from A to X, and then the amplitude for it to go from X to 1. Then these amplitudes must be combined. In this instance the two events, α goes from A to X *then* from X to 1 (we will abbreviate this to $\alpha : A \rightarrow X$, $X \rightarrow 1$), must happen *in succession*, so the amplitudes *multiply*:

$$\text{Amp}(\alpha : A \rightarrow 1) = \text{Amp}(\alpha : A \rightarrow X) \times \text{Amp}(\alpha : X \rightarrow 1)$$

similarly

$$\text{Amp}(\beta : B \rightarrow 2) = \text{Amp}(\beta : B \rightarrow X) \times \text{Amp}(\beta : X \rightarrow 2).$$

This follows from the rules of ordinary probability. If four horses run in a race the chances of randomly selecting the winner are 1/4. If a second race contains 7 horses, then the chances of betting on the winner in that race are 1/7. In total the chance of betting on the winner in both races is $1/4 \times 1/7$ or 1/28.[5]

Now, if we were calculating the probability of a nucleus ending up at detector 1 and another nucleus at detector 2 (we don't mind which one goes where), then according to the normal rules of probability we would say that:

prob(a nucleus at 1 + a nucleus at 2) = prob(nucleus at 1)
$$+ \text{prob(nucleus at 2)}.$$

However, we might suspect that quantum mechanics uses a different rule for this. It turns out, however that this *does* give us the correct result. It is quite complicated in terms of our amplitudes, as we do not mind which nucleus goes where:

prob(a nucleus at 1 + a nucleus at 2)
$$= |\text{Amp}(\alpha : A \rightarrow 1) \times \text{Amp}(\beta : B \rightarrow 2)|^2$$
$$+ |\text{Amp}(\alpha : A \rightarrow 2) \times \text{Amp}(\beta : B \rightarrow 1)|^2$$
$$= 2p$$

(if the probability to go to a detector is the same for each nucleus $= p$).

In this case we do not add all the amplitudes first before we square as the two events we are considering are *distinguishable*. Nucleus α and nucleus β are different, so we can tell in principle (even if we choose not to bother) which one arrives at which detector.

Now we consider a different version of the experiment. This time we arrange that the two nuclei are exactly the same type (say both helium nuclei).

According to normal probability rules it should not make any difference to our results. Yet it does. If we now measure the probability of a nucleus arriving at 1 and another arriving at 2, we get a different answer!

This time, we cannot tell—*even in principle*—if it is the nucleus from A that arrives at 1 or the nucleus from B, so we have to combine the amplitudes *before* we square:

$$\text{prob(a nucleus at 1 + a nucleus at 2)}$$
$$= |\text{Amp}(\alpha : A \rightarrow 1) \times \text{Amp}(\beta : B \rightarrow 2))$$
$$+ (\text{Amp}(\alpha : A \rightarrow 2) \times \text{Amp}(\beta : B \rightarrow 1)|^2$$
$$= |p + p|^2 = |2p|^2$$
$$= 4p$$

a result that is confirmed by doing the experiment.

Classically these two experiments would yield the same result. Classically we would expect that two nuclei will scatter off each other and be detected in the same manner, their being identical nuclei should not make a difference[6].

Consider this analogy. Imaging that you have a bag that contains four balls, two of which are white and two of which are black. The balls are numbered—1 and 2 are black, 3 and 4 are white—so you can tell which balls are being picked out. If you pick out a ball, replace it and then pick out another, the following possibilities are all equally likely:

1B + 1B	1B + 2B	<u>1B + 3W</u>	<u>1B + 4W</u>
2B + 2B	2B + 1B	<u>2B + 3W</u>	<u>2B + 4W</u>
3W + 3W	<u>3W + 1B</u>	3W + 2B	3W + 4W
4W + 4W	<u>4W + 1B</u>	<u>4W + 2B</u>	4W + 3W

which gives the probability of picking out a white ball and a black ball (it doesn't matter which ones—all the underlined are possibilities) as being 8/16 or 50:50.

If we repeat the experiment having wiped off the numbers what are the chances of picking out a white and black ball? Classically they are the same (and indeed for balls they *are* the same as the quantum effects are small on this scale). One cannot imagine that wiping off the numbers would make a difference. However, according to quantum mechanics it does. Now we cannot tell which ball is which and the possible results are:

$$W + W \quad W + B \quad B + B$$

giving a probability of picking out a white and black as 1/3. In other words, if these were quantum balls then the sequences underlined in the first list would all be the same event. It is not simply that we could not *tell* which event we had actually seen, *they are the same event.* Rubbing the numbers off classical balls may make it harder to distinguish them, but it is still possible (using microscopic analysis for example). With the quantum balls the rules seem to be telling us that their identities merge.

Summarizing, the rules for combining amplitudes are:

> if the events cannot be distinguished, amplitudes add
> before squaring;
> if the events can be distinguished, square the amplitudes then add;
> if the events happen in a sequence, multiply the amplitudes.

This relates to the calculation of path amplitudes in the double slot experiment. With the equipment used in the experiment, there is no way to tell which path the electron passes along, so they must all be treated as indistinguishable and their amplitudes added.

3.5   How to calculate with amplitudes

In this section the mathematical rules for combining amplitudes are discussed. Readers with a mathematical bent will gain some insight from this section. Readers without a taste for such matters will not lose out when it comes to the later sections, provided they read the summary at the end of this section.

Amplitudes are not numbers in the ordinary sense, which is why they cannot be measured. Ordinary numbers represent quantities that can be counted (5 apples on a tree, 70 equivalent kilogram masses in a person, 1.6×10^{-19} coulombs charge on a proton). Ordinary numbers have size. Amplitudes have both *size* and *phase*.

Consider a clock face (figure 3.6). The short hand on a clock has a certain length (size). At any moment it is pointing to a particular part of the clock face. By convention, we record the position of the hand by the angle that it has turned to from the straight up (12 o'clock) position.

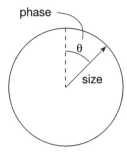

Figure 3.6 The size and phase of a clock hand.

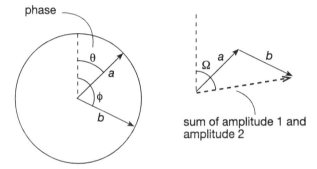

sum of amplitude 1 and
amplitude 2

Figure 3.7 The rule for adding amplitudes is to place them nose to tail
the total amplitude is the line connecting the first tail to the second nose.

To tell the time we must specify the size (length) and phase (angle) of
both hands on the face.

The rule for adding quantities that have both size and phase was
established by mathematicians long before quantum mechanics was
developed (figure 3.7):

---

amplitude 1: size a, phase θ

amplitude 2: size b, phase ϕ

amplitude 3 = amplitude 1 + amplitude 2

amplitude 3 has size r where $r^2 = a^2 + b^2 + 2ab\cos(\phi - \theta)$

and phase Ω is given by $r\sin(\Omega) = a\sin(\theta) + b\sin(\phi)$.

These are quite complicated rules and we will not be using them in this book. However, note that according to the rule for adding sizes, two amplitudes of the same size can add to make an amplitude of zero size if their phases are different by 180° (figure 3.8).

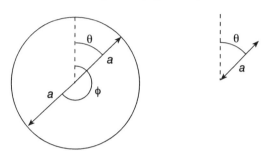

Figure 3.8 These two amplitudes have the same size, but the phases are 180° different when they are added together, so the total is zero.

If amplitude 1 has size a, and phase θ and amplitude 2 has size a, and phase $\phi = 180° + \theta$, then amplitude 3 has size r where

$$
\begin{aligned}
r^2 &= a^2 + b^2 + 2ab\cos(\phi - \theta) \\
&= r^2 + r^2 + 2rr\cos(180°) \\
&= 2r^2 + 2r^2(-1) \\
&= 0 \qquad (\text{as } \cos(180°) = -1).
\end{aligned}
$$

There are also rules for multiplying amplitudes together:

amplitude 1 × amplitude 2 = amplitude 3
size of amplitude 3 = $(a \times b)$
phase of amplitude 3 = $\theta + \phi$.

The comparative simplicity of these rules suggests that multiplying is a more 'natural' thing to do with amplitudes than adding them.

If the two amplitudes have the same size, but phases that are opposite (e.g. 30° and −30°) then when they are multiplied together the size is squared ($a \times a = a^2$) but the phase is zero. Such pairs of amplitudes are called *conjugates*, and multiplying an amplitude by its conjugate is *absolute squaring* the amplitude. Whenever an amplitude is squared in this fashion, the phase is always zero. Our earlier rule relating amplitudes to probabilities should more properly read:

$$\text{probability} = \text{amplitude} \times \text{conjugate amplitude} = |\text{amplitude}|^2$$

which is why we never have to deal with the phase of a probability—it is always zero.

Summary

- Amplitudes have both size and phase;
- phase is measured in degrees or radians;
- adding amplitudes is a complex process which results in two amplitudes of the same size adding up to zero if they are 180° different in phase;
- when amplitudes are multiplied together their sizes multiply and their phases add;
- conjugate amplitudes have the same size but opposite phase;
- multiplying an amplitude by its conjugate produces a quantity with a phase of zero—an ordinary number.

3.6 Following amplitudes along paths

A path consists of a set of points and a list of times (see figure 3.9). A particle moving along the path will arrive at a given point at a given time. If the path is to be completely specified then an infinite number of points and times are needed. However, a good approximation can be made by considering a finite, but large, number of points.

To approximate the amplitude for a given path Feynman laid down the following rules:

- every path contributes an amplitude with the *same size*, but a *different phase*;
- the amplitude along the whole path can be found by multiplying together amplitudes for each part of the path.

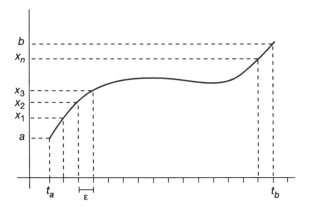

Figure 3.9 A path (here one-dimensional) can be split up into a series of points by specifying times and the position at those times. In this case the time has been split up into *n* equal intervals ε along the horizontal time axis.

So, along the path shown:

$$A(a, b) = A(a, x_1) \times A(x_1, x_2) \times A(x_2, x_3) \times \cdots \times A(x_n, b).$$

According to the rules quoted in the last section, when amplitudes are multiplied together their sizes *multiply* and their phases *add*, so:

$$\text{size of } A(a, b) = (\text{size of each amplitude})^n$$

$$\text{phase of } A(a, b) = \text{sum of phases for each amplitude}$$

$$\text{along the path.}$$

So far we have been dealing with the rules for combining amplitudes— the physics lies in the calculation of the phase. Feynman discovered that the phase for an amplitude to get from one point on a path to the next was given by:

$$\text{phase}(x_1, x_2) = \frac{\varepsilon}{\hbar} L$$

in which $\varepsilon = t_2 - t_1$ is the small time interval between the two points on the path, $\hbar$ is Planck's constant h divided by 2π, and L is the *Lagrangian* of the system defined by:

$$L = \frac{1}{2}mv^2 - V(x, t) = \frac{1}{2}m\frac{(x_2 - x_1)^2}{\varepsilon} - V\left(\frac{x_1 + x_2}{2}, \frac{t_2 + t_1}{2}\right)$$

i.e. the difference between the kinetic energy of the particle (for particles not moving near to the speed of light this is $\frac{1}{2}mv^2$) and the potential energy V calculated at a point midway between the start and end points. The Lagrangian is a well-known mathematical object from classical physics, but Feynman was the first person to see its relevance to quantum mechanics[7].

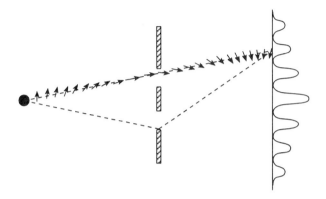

Figure 3.10 The phase (represented by the arrows) advances along the path.

As we move from one point to the next along a path the phase smoothly changes in proportion to the time interval and the Lagrangian. Figure 3.10 illustrates this for a path in the double slot experiment. In this instance, as the path arrives at one of the peaks in the distribution, many of the paths arriving at the same point must arrive with the same phase so they add up. Many other paths may arrive at the point with very different phases, but enough of them must be different by 180° to cancel each other out leaving an overall amplitude. The points where no electrons arrive correspond to paths arriving that have amplitudes that all cancel each other out.

3.7 Relating quantum mechanics to the everyday world

If we summarize Feynman's prescription for quantum mechanics:

- every path contributes an amplitude;

- the total amplitude is found by adding the amplitudes for each path;
- each path can be split up into pieces and the amplitude for each piece calculated and then multiplied together to get the amplitude for the whole path;
- each amplitude has the same size, but the phase depends on the Lagrangian;
- some paths arrive with similar phases, so they add to increase the total amplitude;
- some paths have amplitudes with phases that are 180° or more apart, so they add and reduce the total amplitude;
- the probability is found by absolute squaring the total amplitude.

The rules are quite clear that *every* path must be considered. If this is not done, then the calculated probability (as checked by experiment) is wrong. This means that *even paths that are impossible judged by classical physics must be taken into account.*

All of this leaves us with a very big problem. How can we relate the behaviour of objects that we use every day to this weird behaviour? Are we to believe that each electron follows every path at once? If not, why must each contribution be included? If in some sense they do, then how come when I throw a cricket ball to my eldest son, I only see it following one path (one can't even imagine what it would mean to have the ball follow all possible, and impossible, paths at once). What can we distill from the mathematics about the actual way in which particles such as electrons behave?

These are all very difficult questions and we have not yet found a way of answering them completely. There are several different partial answers each of which is supported by various physicists. It is difficult to be completely sure what Feynman's view on this was as he was not one to philosophize. He certainly believed that physics was all about calculating results and comparing them to experiment. In some sense he did not mind if the mathematics produced results that defied common sense—after all he would say this is how nature *is*, and if we don't like the idea then its our problem not nature's!

However, we do not need to leave it just at that point as Feynman's way of calculating does give some further insight. When you throw a ball you see it follow one path—the classical path of the ball. It turns out that in quantum mechanics the phase of the amplitude along any path 'close' to the classical path is almost identical to the phase of the classical path's amplitude. This is very important as it implies that the amplitudes for paths close to the classical one will all tend to add to one another. This is not generally true for any set of 'close' paths, only ones close to the classical path.

Now, at this point the size of the object becomes important. Between any two points on the ball's path there is a very large distance from the point of view of the electrons and atoms within the ball. So, every two points on the classical path are connected by an enormous number of quantum paths.

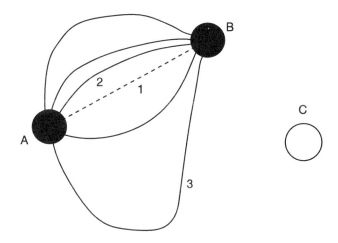

Figure 3.11 The classical path of a ball can be broken into points, such as A and B, connected by many quantum paths.

In figure 3.11, A and B represent two points on the classical path of a ball. They have been connected by some different possible paths. The dashed line is the exact classical path. Some other paths close to this one (such as 2) have been added. Their amplitudes will have phases that will be almost the same as that along the dashed path. Each of these close paths will add to the amplitude. Paths further from the classical, such as 3, will have very different phases and will tend to cancel each other out.

As the value of $\hbar$ is very small (1.05×10^{-34} J s) and the value of the Lagrangian for a classical object is comparatively large, a small change in the Lagrangian along any path makes a big difference to the phase. Paths that are not *very* close to the classical will have quite different phases. Consequently only a few paths close to the classical actually contribute to the amplitude. All the others tend to cancel each other out.

If we calculate the amplitude for the ball to go off the classical path, to some point such as C for example, then *all* the paths leading to this point will be way off classical. The amplitude in total will be very small as all the paths leading to it cancel each other out. We never see the ball following a non-classical path as the probability of it getting to some point such as C is vanishingly small. All the quantum excursions along odd paths all at once are confined to tiny regions between points on the classical path.

What this does do, however, is destroy the classical view that the ball moves from one point to another continuously along the classical path. The motion is far more complex than this and what we see is a sort of 'smeared out' average.

3.8   Energy levels

It is quite commonly known that electrons inside atoms can only exist with certain values of energy called the *energy levels*. If you try to insert an electron into an atom with a different energy, then it will not fit and either come out of the atom, or radiate away the difference in energy and settle into one of the levels. This is often compared to the rungs on a ladder. One can stand on the rungs, but if you try to stand between the rungs all you do is fall down onto the next level. Energy levels are a common quantum mechanical consequence of having a closed system, i.e. a system where the paths are localized in space.

Electrons within an atom must stay within the boundary of the atom as they do not have enough energy to escape. If we consider the motion of electrons within atoms as being closed paths (i.e. ones that return to their starting point) then we can calculate the amplitude to go round the atom and back to the start. If we do this, then the phase at the start of the path should be the same as the phase at the end. This must be so as the start and end are the same place and even quantum mechanics does not allow two different phases at the same point! Within a given atom, the Lagrangian is such that only certain energy values allow this to happen. Hence there are energy levels within atoms.

The same is true of other closed system. Quarks within hadrons have energy levels; however, in this case the Lagrangian is much more complicated as the strong force between the quarks has to be taken into account when calculating the potential energy.

3.9 Photons and waves

In the previous chapter I mentioned that photons, which in many ways seem like particles, can be assigned a wavelength which is a property traditionally only associated with waves. This is another aspect of the quantum nature of objects. The double slot experiment for electrons can also be carried out for photons with exactly the same results (see the quotation from Feynman at the start of this chapter). When these experiments were first carried out they seemed to be indicating that both electrons and photons could behave like particles in some circumstances (i.e. with only one slot open) and like waves in another (both slots open, and the objects passing through both at the same time).

The double slot experiment was first carried out for light by Young in 1801 (see figure 3.12). It was interpreted as evidence for the wave nature of light.

When light waves from a small source strike the screen they produce secondary waves centred on the slots. These waves will then arrive at the photographic film having travelled from one slot or the other. Depending on the distance travelled the waves will either arrive in phase with each other (i.e. the peaks or dips on the waves arrive at the same moment) or out of phase (i.e. a dip on one wave arrives with a peak from another). In phase produces re-enforcement and hence a patch of light on the film, out of phase produces cancellation and so a dark region.

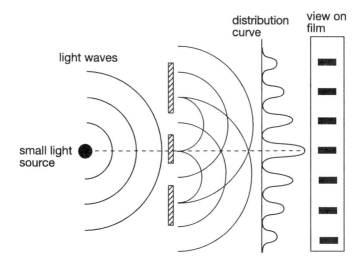

Figure 3.12 A double slot experiment for light.

However, if this experiment is done in a more sophisticated manner using a laser, then a puzzling aspect emerges.

If the intensity of the laser is turned right down and the film replaced with electronic detectors, then the pattern of light and dark bands beaks up into a pattern of speckles, just like the pattern for individual electrons. This is interpreted as evidence for the laser producing individual photons that cross the experiment one at a time. Yet, just like the electrons they 'know' about both slots being open.

This is the crux of one of the most bizarre aspects of quantum mechanics. Wave and particle natures seem interchangeable.

Feynman's sum over paths approach to quantum mechanics gives us another view. Electrons and photons are neither waves nor particles. They are quantum objects behaving in a manner given by the amplitudes. Amplitudes behave like waves (they have a size and a phase). Both photons and electrons can be given an equivalent wavelength (figure 3.13) by considering the progression of the phase of the amplitude as they move.

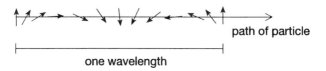

Figure 3.13 The wavelength of an amplitude.

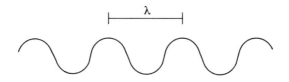

Figure 3.14 The wavelength, λ, of a wave.

The distance between points along the path of the particle at which the phases are equal is the wavelength of the particle. This is identical to the definition of the wavelength of a wave (figure 3.14).

Likewise, the frequency of a particle can be defined as the number of complete rotations of the phase in one second.

In the case of a photon, the Lagrangian contains the relativistic energy and no potential energy while it is moving through space, i.e.:

$$L = pc$$

$$\therefore \qquad \text{phase} = \frac{pct}{\hbar} = \frac{px}{\hbar}$$

where x is the distance travelled by the photon. One complete revolution of the phase is 360° or 2π radians, so the distance between points of equal phase is given by:

$$2\pi = \frac{px}{\hbar}$$

$$\therefore \qquad x = \frac{\hbar \times 2\pi}{p} = \frac{h}{p}$$

and calling this distance λ, we have:

$$p = \frac{h}{\lambda}$$

which is exactly the relationship we produced in the previous chapter (equation (2.15)).

It pays to calculate some wavelengths to see the sort of distances that we are dealing with in different circumstances:

$$p = \frac{h}{\lambda}$$

for an electron, with $p \sim 60$ keV/c

$$\lambda = \frac{h}{p} = \frac{6.63 \times 10^{-34}}{3.2 \times 10^{-23}} = 2.1 \times 10^{-11} \text{ m}$$

which is comparable to atomic sizes. On the other hand, a cricket ball travelling at 10 m s^{-1} would have a quantum wavelength of:

$$\lambda = \frac{h}{p} = \frac{6.63 \times 10^{-34}}{0.1 \text{ kg} \times 10 \text{ m s}^{-1}} = 6.63 \times 10^{-34} \text{ m}$$

which is far too small to be observed.

3.10   Summary of chapter 3

- The ordinary rules of probability do not apply in the microscopic world of atoms and sub-atomic particles;
- quantum theory is a new way of calculating that was developed to cope with the strange results of atomic experiments;
- in quantum mechanics every event has an amplitude which is absolute squared to give the probability of the event happening;
- amplitudes can be added and multiplied to account for alternative 'paths' to the event;
- amplitudes sometimes combine to increase the overall amplitude and sometimes they combine to decrease the overall amplitude;
- Feynman's rules for calculating amplitudes state that every path to an event must be considered, no matter how far it is from the common sense 'classical' path;
- the key to calculating the amplitude of a path is the Lagrangian of the particle moving along that path;

- the 'real' world is the quantum world, what we see in our large-scale world is an average of the quantum effects that are actually going on all the time;
- photons behave in a similar manner to electrons;
- photons and electrons can be given a wavelength and a frequency based on rotation of the phase of the amplitude along a path.

Notes

[1] Einstein said that common sense was a prejudice laid down before the age of 18.

[2] Electron double slit experiments are not easy to do! The first such experiment was performed by G Mollenstedt and H Duker in 1954. These days they are done with beams from an electron microscope which typically have energies of 60 keV. The beam is aimed at a pair of slits about 0.5 μm wide separated by 2 μm. With the detector placed about 2 m from the slits the distance between dark lines on the pattern is 5 μm.

[3] Although I have not suggested any mechanism that enables this 'communication' between electrons, it is important to eliminate it as a possibility.

[4] Undergraduates learn the 'older' view of quantum theory first as they need to develop their mathematical skills before they can go on to Feynman's view. Often they will not see the later way of doing calculations unless they go on to graduate work in the subject.

[5] Apply this to the UK lottery. Each lottery ball has 49 numbers and 6 numbers need to be correct. The chances of selecting the right sequence are $(1/49)^6$. Not good odds!

[6] Barring such effects as the difference in mass and size etc, which can be compensated for in the calculations.

[7] He was following up on a suggestion made by Dirac.

Chapter 4

The leptons

In this chapter we shall study the leptons and introduce the way in which we describe particle reactions. In the process we shall learn how conservation of electrical charge applies to particle physics. The solar neutrino problem will be discussed and some features of the weak force will become apparent.

4.1 A spotter's guide to the leptons

Leptons, unlike quarks, can exist as separate objects (remember that the strong force binds quarks into particles, so that we can never study a single quark without its partners). Hence their reactions are more straightforward than those of the quarks. This makes it easier to develop some of the ideas of particle physics using leptons as examples. Later we can apply these ideas to the slightly more complicated case of the quarks.

Particle physics can seem like a very 'unreal' subject which studies rare and unimportant sounding objects. This is not the case. Table 4.1 shows that some of the leptons are readily found in nature and have important roles to play in common processes.

The electron is a well-known particle and its properties are established in basic physics courses. Its partner in the first generation, the electron-neutrino, is less well known but just as common in nature. Some radioactive processes produce them and the central cores of atomic

Table 4.1 Where to find leptons.

1st generation	2nd generation	3rd generation
electron	muon	tau
(i) found in atoms (ii) important in electrical currents (iii) produced in beta radioactivity	(i) produced in large numbers in the upper atmosphere by cosmic rays	(i) so far only seen in labs
electron-neutrino	muon-neutrino	tau-neutrino
(i) produced in beta radioactivity (ii) produced in large numbers by atomic reactors (iii) produced in huge numbers by nuclear reactions in the sun	(i) produced by atomic reactors (ii) produced in upper atmosphere by cosmic rays	(i) so far only seen in labs

reactors emit them in very large numbers. The sun is an even more copious producer of electron-neutrinos. Approximately 10^{12} electron-neutrinos pass through your body every second, the vast majority of which were produced by the nuclear reactions taking place in sun's core. Fortunately, neutrinos interact with matter so rarely that this huge number passing through us does no harm whatsoever.

The members of the second generation are less common, but they are still seen frequently enough in nature to make their study relatively straightforward.

Muons are easily produced in laboratory experiments. They are similar to electrons, but because they are more massive they are unstable and decay into electrons and neutrinos (in just the way an unstable atom will decay radioactively). Unlike electrons they are not produced by radioactive decay. They are easily observed in cosmic ray experiments.

The members of the third generation have not been seen in any naturally occurring processes—at least not in the current era of the universe's evolution. In much earlier times, when the universe was hotter and particles had far more energy, the members of the third lepton generation would have been produced frequently in particle reactions.

This, however, was several billion years ago. Today we only observe the tau in experiments conducted by particle physicists. The tau-neutrino has not been observed directly, but it presence can be inferred from certain reactions.

4.2   The physical properties of the leptons

Table 4.2 reproduces the standard generation representation of the leptons.

Table 4.2 The leptons (M_p = proton mass). Charge of particles is -1 in the top row and 0 in the bottom row.

1st generation	2nd generation	3rd generation
electron	muon	tau
$5.45 \times 10^{-4} M_p$ (5.11×10^{-4} GeV/c^2)	$0.113 M_p$ (0.106 GeV/c^2)	$1.90 M_p$ (1.78 GeV/c^2)
electron-neutrino	muon-neutrino	tau-neutrino
$< 1.07 \times 10^{-8} M_p$ ($< 1.14 \times 10^{-8}$ GeV/c^2)	$< 2.67 \times 10^{-4} M_p$ ($< 2.85 \times 10^{-4}$ GeV/c^2)	$0.075\ M_p$ (< 0.080 GeV/c^2)

The electron, muon and tau all have an electrical charge of -1 on our standard scale, while the three neutrinos do not have an electrical charge.

Table 4.2 also shows that the charged leptons increase in mass from generation to generation. However, the neutrino masses are not so clear cut.

Our best experimental measurements have been unable to establish a definite mass for any of the neutrinos. The numbers in table 4.2 are the upper limits that we are able to place on the various neutrino masses. The increase in these numbers across the table displays our increasing *ignorance* of the masses, not necessarily an increase in the *actual* masses of the neutrinos. It is not possible to 'capture' just one example of a particle and place it on a set of scales. Masses have to be calculated by studying the ways in which the particles react. The more particles one measures, the more precise one can be about the size of their mass.

For example, the tau-neutrino is a very rare object that is difficult to study. All we can say at this time is that its mass (if it has one) is less than 0.075 proton masses. This limit comes from studying the decay of the tau, which is itself a rare object. The uncertainty will become smaller as we measure more of them. The electron-neutrino, on the other hand, is much more common, so the upper limit on its mass is much more clearly defined.

Unfortunately, our theories do not predict the masses of the neutrinos. For a long time particle physicists assumed that neutrinos have no mass at all, but recently some measurements have suggested that they may have an extremely small, but finite, mass. This is an open question in experimental physics. The 'standard version' of the standard model assumes that the neutrinos have no mass. If they do, then the theories can be adapted to include this possibility—indeed, as we shall see, it might even solve an interesting problem in the formation of galaxies (see chapter 13).

Neutrinos seem like very weird particles. They may have no mass (although professional opinion is swinging towards their having *some* mass) and they certainly have no charge, so they are very ghostly objects. Being leptons they do not feel the strong force and having no electrical charge they do not feel the electromagnetic force either. The only way that neutrinos can interact is via the weak force. (Ignoring gravity, which would be extremely weak for an individual neutrino. However, if there are enough neutrinos with mass in the universe, their *combined* gravity might stop the expansion of the universe.)

As weak force reactions are very rare, a particle that *only* reacts by the weak force cannot be expected to do very much. Estimates indicate that a block of lead *ninety thousand million million* metres thick would be required to significantly reduce the number of electron-neutrinos passing through your body at this moment.

4.3   Neutrino reactions with matter

Neutrinos are difficult to use in experiments, but they are worth the effort for one reason: they are the only particles that allow a relatively uncluttered study of the weak force. After all, they do not react in any other way. Their reactions may be very rare, but if you have enough neutrinos to begin with then sufficient reactions can take place for a sensible study to be made.

A characteristic reaction involving a neutrino is symbolized in equation (4.1):

$$\nu_e + n \rightarrow p + e^-. \tag{4.1}$$

This reaction has been of considerable experimental importance over the past few years. Astrophysicists have calculated the rate at which electron-neutrinos should be being produced by the sun. In order to check this theory physicists have, in several experiments, measured the number of neutrinos passing through the earth. The first of these experiments relied on the neutrino–neutron reaction (4.1).

If the neutron in (4.1) is part of the nucleus of an atom, then the reaction will transform the nucleus. For example, a chlorine nucleus containing 17 protons would have one of its neutrons turned into a proton by the reaction, changing the number of protons to 18 and the nucleus to that of argon:

$$^{37}_{17}\text{Cl} + \nu_e \rightarrow {}^{37}_{18}\text{Ar} + e^-. \tag{4.2}$$

The electron produced by the reaction would be moving too quickly to be captured by the argon nucleus. It would escape into the surrounding material. However, the transformation of a chlorine nucleus into an argon nucleus can be detected, even if it only takes place very rarely. All one needs is a sufficiently large number of chlorine atoms to start with.

The Brookhaven National Laboratory ran an experiment using 380000 litres of chlorine-rich cleaning fluid in a large tank buried in a gold mine many metres below the ground. Few particles can penetrate so far underground and enter the tank. Mostly these will be the weakly reacting solar neutrinos. To isolate the equipment from particles produced in the rock, the tank was shielded. The energy of solar neutrinos is sufficient to convert a nucleus of the stable isotope chlorine 37 into a nucleus of argon 37. Argon 37 is radioactive. Every few months the tank was flushed out and the argon atoms chemically separated from the chlorine. The number of argon atoms was then counted by measuring the intensity of the radiation that they produced.

Theoretical calculations show that the number of reactions should be approximately 8 per second for every 10^{36} chlorine 37 atoms present in the tank. The experiment ran from 1968 to 1986 and over that time measured only 2 per second for every 10^{36} atoms in the tank. Both numbers represent incredibly small rates of reactions (a measure of how weak the weak force is) but the conclusion is clear—the measured number of electron-neutrinos produced from the sun is totally different from the theoretical predictions. This is the solar neutrino problem. It is currently unresolved.

This is an example of a very concrete piece of physics based on the properties of these ghostly particles, and a genuine mystery in current research.

4.3.1 Aspects of the neutrino–neutrino reaction

4.3.1.1 *Conservation of electrical charge*

Returning to the basic reaction (4.1) within the chlorine nucleus:

$$\nu_e + n \rightarrow p + e^- \qquad (4.1)$$

the only way that a neutron can be turned into a proton is for one of the d quarks to be turned into a u. Evidently, inside the neutron, the following reaction has taken place:

$$\nu_e + d \rightarrow u + e^-. \qquad (4.3)$$

This is the basic reaction that has been triggered by the weak force. Notice that it only involves one of the quarks within the original neutron. The other two quarks are unaffected. This is due to the short range of the weak force. The neutrino has to enter the neutron and pass close to one of the quarks within it for the reaction to be triggered.

This is quite a dramatic transformation. The up and down quarks are different particles with different electrical charges. Examining the reaction more carefully:

$$\nu_e \quad + \quad d \quad \rightarrow \quad u \quad + \quad e^-$$
$$\text{charges} \quad 0 \quad \quad -\tfrac{1}{3} \quad \quad +\tfrac{2}{3} \quad \quad -1$$

we see that the individual electrical charges of the particles involved have changed considerably. However, if we look at the *total* charge involved then we see that the total is the same after the reaction as before. On the left-hand side the charge to start with was $(0)+(-1/3) = -1/3$, and on the right the final total charge of the two particles is $(+2/3) + (-1) = -1/3$ again.

This is an example of an important rule in particle physics:

CONSERVATION OF ELECTRICAL CHARGE
In any reaction the total charge of all the particles entering the reaction must be the same as the total charge of all the particles after the reaction.

Although we have illustrated this rule by the example of a weak force reaction involving neutrinos the rule is much more general than that. Any reaction that can take place in nature must follow this rule, no matter what force is responsible for the reaction. It is good practice to check the rule in every reaction that we study.

4.3.1.2 *Muon neutrinos and electron-neutrinos*

There is an equivalent reaction to that of the electron-neutrino involving muon-neutrinos:

$$\nu_\mu + n \rightarrow p + \mu^-. \tag{4.4}$$

In this case the neutrino has struck the neutron transforming it into a proton with the emission of a muon. The incoming muon-neutrino has to have more energy than the electron-neutrino as the muon is 200 times more massive than the electron. The muon will be produced with a large velocity and will be lost in the surrounding material. The reaction is very similar to the previous one. At the quark level:

$$\nu_\mu + d \rightarrow u + \mu^- \tag{4.5}$$

a reaction that also conserves electrical charge.

There is one important contrast between the reactions. In all the reactions of this sort that have ever been observed, the muon-neutrino has *never* produced an electron, and the electron-neutrino has never produced a muon[1].

This is a prime example of the weak force acting *within* the lepton generations. The electron and its neutrino are in the first lepton generation; the muon and its neutrino are in the second generation. The weak force cannot cross the generations. It was the contrast between these two reactions that first enabled physicists to be sure that there were different types of neutrino.

The existence of the neutrino was predicted by the Viennese physicist Wolfgang Pauli[2] in 1932. Unfortunately, it was not until atomic reactors were available in the 1950s that sufficient neutrinos could be generated to allow the existence of the particles to be confirmed experimentally. The large numbers produced by a reactor core were needed to have sufficient reactions to detect[3].

By the late 1950s the existence of the muon had been clearly established (it had first shown up in cosmic ray experiments in 1929) and its similarity to the electron suggested that it might also have an associated neutrino. However, as neutrino experiments are so hard to do, nobody was able to tell if the muon-neutrino and the electron-neutrino were different particles.

The problem was solved by a series of experiments at Brookhaven in the 1960s. A team of physicists, lead by Leon Lederman, managed to produce a beam of muon neutrinos[4]. The neutrino beam was allowed to interact with some target material. Over a 25-day period some

10^{14} muon-neutrinos passed through the experiment. The reactions between the neutrinos and the nuclei in the target produced 51 muons. No electrons were produced. In contrast, electron-neutrinos readily produced electrons when passed through the target material, but no muons. The results of Lederman's experiment were:

$$\nu_\mu + n \rightarrow p + \mu^- \quad \text{51 events}$$

$$\nu_\mu + n \rightarrow p + e^- \quad \text{no events}$$

but

$$\nu_e + n \rightarrow p + e^-$$

so the ν_e and the ν_μ are different! This experiment clearly established the fundamental distinction between these two types of neutrino.

A later series of experiments at CERN found the ratio of electrons to muons produced by muon-neutrinos to be 0.017 ± 0.05, consistent with zero. For his pioneering work in identifying the two types of neutrino Lederman was awarded the Nobel Prize in physics in 1989[5].

Now, of course, we realize that there are three neutrinos, the tau-neutrino being in the third generation of leptons along with the tau.

4.4   Some more reactions involving neutrinos

The fundamental distinction between muon-neutrinos (ν_μ) and the electron-neutrinos (ν_e) is reflected in a variety of reactions. For example, there is an interesting variation on the basic reaction that lies behind the solar neutrino counting experiment.

The basic ν_e reaction converts nuclei of chlorine 37 into radioactive argon 37. It is also possible for the argon 37 atom to reverse this process by 'swallowing' one of its electrons:

$$^{37}_{18}\text{Ar} + e^- \rightarrow ^{37}_{17}\text{Cl} + \nu_e. \quad (4.6)$$

This reaction is called 'K capture' because the electron that triggers the reaction comes from the lowest energy state of the orbital electrons, called the K shell. Electrons in this lowest energy state spend much of their time in orbit very close to the surface of the nucleus. Sometimes they pass *through* the outer layers of the nucleus. With certain atoms, and argon 37 is an example, this electron can be 'swallowed' by one of the protons in the nucleus turning it into a neutron. This process is another example of a reaction triggered by the weak force. Once again the weak force must work within the lepton generations, so the member of the first generation that was present before the reaction, the electron, has to be replaced by another member of the first generation after the reaction—in this case the ν_e that is emitted from the nucleus. At the underlying quark level:

$$u + e^- \rightarrow d + \nu_e \qquad (4.7)$$
$$\tfrac{2}{3} + -1 = -\tfrac{1}{3} + 0.$$

Again we note that charge is conserved in this reaction.

Another reaction that illustrates the weak force staying within generations is the collision of a ν_μ with an e^-:

$$\nu_\mu + e^- \rightarrow \mu^- + \nu_e. \qquad (4.8)$$

This is a very difficult reaction to study. Controlling the neutrino beam and making it collide with a sample of electrons is hard to arrange.

In this reaction a ν_μ has been converted into a μ^-, and an e^- has been converted into a ν_e. Notice that the weak force has again preserved the generation content in the reaction. The particles have changed, but there is still one from the first generation and one from the second on each side of the reaction.

We have now accumulated enough evidence to suggest the following rule:

> **LEPTON CONSERVATION**
> In any reaction, the total number of particles from each lepton generation must be the same before the reaction as after.

All the reactions that we have looked at have followed this rule (go back and check). Physicists regard this as one of the fundamental rules of particle physics. There are two points worth noting about this rule:

- it must also be followed by the other forces, as the weak force is the only one that allows fundamental particles to change type;
- it is not clear how the rule applies if there are no particles from a given generation present initially.

The second point can be illustrated by an example from nuclear physics.

Beta radioactivity takes place when a neutron inside an unstable isotope is transformed into a proton by the emission of an electron. In terms of observed particles the reaction is:

$$n \rightarrow p + e^-. \tag{4.9}$$

The reaction is incomplete as it stands. When β decay was first studied only the electron was observed, which was a puzzle as the reaction did not seem to conserve energy or momentum (see page 146). We can also see that it does not appear to observe the lepton conservation rule.

At the quark level there is a fundamental reaction taking place:

$$d \rightarrow u + e^-. \tag{4.10}$$

As a quark has changing type, we must suspect that the weak force is involved[6]. However, on the left-hand side of the equation there is no member of the first generation present. On the right-hand side an electron has appeared—so now there is a first generation representation. This is an odd situation. Either the rule is violated in certain circumstances, or we have not yet found out how to consistently apply the rule. In the next chapter we shall explore this in more detail and discover how we can make the rule *always* apply.

4.5 'Who ordered that?'[7]

So far we have concentrated on the massless (or at least very low mass) particles in the lepton family. We have not really discussed the reactions and properties of the massive leptons. In part this is because in broad terms they behave exactly like electrons with a greater mass.

For example, muons are common particles that are easily produced in reactions. They have a negative charge which all measurements indicate is *exactly* the same as that of the electron. The only apparent difference between muons and electrons is that the muon is 200 times more massive. We have no idea what makes the muon so massive.

The muon was first discovered in experiments on cosmic rays[8]. Cosmic rays are a stream of particles that hit the earth from the sun and other stars in the galaxy. 75% of the cosmic rays counted at sea level are muons. These muons, however, do not originate in outer space. They are produced high up in the earth's atmosphere. Protons from the sun strike atoms in the atmosphere causing reactions that produce muons[9]. We can be quite sure that the muons we observe at sea level have not come from outer space because the muon is an unstable particle.

It is impossible to predict when a given muon will decay (just as it is impossible to predict when a given atom will decay), but we can say on average how long a muon will live for. Particle physicists call this the *lifetime* of the particle.

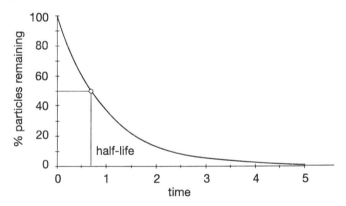

Figure 4.1 The decay of particles.

Figure 4.1 shows how the number of particles decreases with time in a batch of identical particles that were all created at the same moment. The same shape of curve would be produced if the number of radioactive atoms were plotted instead. The half-life is the amount of time taken to reduce the number of particles (or atoms) by 50%. The lifetime is the average life of a particle in the set.

Muon lifetimes have been measured very accurately; the quoted value is $(2.19703 \pm 0.0004) \times 10^{-6}$ seconds (i.e. about 2 millionths of a second[10]). This seems a remarkably short length of time, but compared to some the muon is a very long lived particle. This is why the muons *must* be produced in the upper atmosphere. They do not live long enough to reach the surface of the earth from any further away. Indeed, most muons produced in the upper atmosphere decay before they reach sea level. That any muons manage to survive is a direct consequence of Einstein's theory of relativity which predicts that time passes more slowly for moving objects[11].

The physics of decay processes is worthy of a chapter on its own (chapter 8) but we can briefly discuss one important point here: the particles produced by the decay *did not exist before the decay took place*.

This is quite difficult to imagine. The muon, for example, does not consist of an electron and some neutrinos bound together. As we have been saying for some time, the muon is a fundamental particle and so does not have any objects inside it. The decay process is not like a 'bag' splitting open to release the electron and neutrinos that were held inside. At the moment of decay the muon disappears and is replaced by the electron and the neutrinos. This seemingly magical process is fully explained by Einstein's theory of relativity.

If you have difficulty believing that such a process can take place, then consider the emission of light by electrons. Atoms emit light when one of their electrons drops from a high energy level to a lower level. Ask yourself if the light wave (or photon) existed *before* the electron dropped levels. Would it be true to say that the electron was composed of an electron and a light wave before the emission?

Decay processes and the emission of light are both examples of the transformation of energy from one type to another. The electron loses energy as it drops a level and this energy is emitted in the form of a light wave (or photon). Similarly, the energy locked up in the muon is turned into a different form—the electron and neutrinos. This turning of energy into matter is a consequence of $E = mc^2$.

The tau is also unstable. It can decay, for example, into a muon and neutrinos (the muon will in turn decay as well). Being more massive than the muon (about 17 times) the tau has an even smaller lifetime of $(3.3 \pm 0.4) \times 10^{-13}$ seconds. The tau is not as well studied as the other leptons. It is a comparatively recent discovery (1975) and difficult to produce in large numbers.

The decays of the muon and tau raise another point about the lepton conservation rule:

$$\mu^- \rightarrow e^- + \nu_\mu + \nu_? \tag{4.11}$$

$$\tau^- \rightarrow \mu^- + \nu_\tau + \nu_?. \tag{4.12}$$

We can deduce the presence of the ν_μ in (4.11) from the rule of lepton conservation. We started with a member of the second generation (μ^-), so we must end up with a member of the second generation (ν_μ). The presence of a second neutrino among the decay products is not so obvious. How can we be sure that it is there?

The existence of the second neutrino can be deduced by measuring the energy of the electrons. The result shows that the electrons vary considerably in energy from one decay to another. If there was only one neutrino, then there would only be one way of sharing the energy out between the two particles. Hence the electrons would always have the same energy. If there are three particles, then there are an infinite number of ways of sharing the energy among the three—hence we can find the electron with a spread of energies[12].

We have argued that one of these neutrinos must be a ν_μ on the grounds of lepton conservation, but what of the other? As it stands the reaction does not conserve leptons overall. The number of second generation particles balances, but there is a first generation particle among the decay products (the e^-) and there was no such particle initially.

This is the same problem that we saw in the context of neutron decay. The problem is repeated in the tau decay equation (4.12). The rule does not seem to work as we understand it at the moment. We need to develop a more complete understanding in the next chapter.

4.6 Summary of chapter 4

•   The electron, the muon and their neutrinos are very common fundamental particles in the universe;

- the tau is too heavy to be produced naturally at this stage of the universe's evolution, but was much more common billions of years ago;
- the tau-neutrino is not common today because the tau is not produced;
- all neutrinos appear to have zero mass;
- neutrinos can react with matter, but very rarely, so experiments have to use large numbers of them;
- experiments to measure the number of electron-neutrinos from the sun are currently finding far fewer than theory would suggest;
- *all reactions follow the rule of conservation of electrical charge*;
- the weak interaction only works within lepton generations.

Notes

[1] Even if the electron-neutrino had sufficient energy to create the muon.

[2] Pauli is well known for the Pauli principle—that electrons in atoms cannot sit in exactly the same state as each other—and the Pauli effect—that any equipment would spontaneously go wrong in his presence. Pauli was a *theoretical* physicist.

[3] The experiment was performed by a team lead by F Reines between 1953 and 1956. By the end of this time they were able to confirm that of the 10^{20} neutrinos produced by the reactor every second, one reacted with their equipment every twenty minutes. Pauli died in 1958, his prediction confirmed after a 24-year wait.

[4] They used the decay $\pi^+ \rightarrow \mu^+ + \nu_\mu$ and filtered out the muons by passing them through lead sheets.

[5] There is a well known story regarding Lederman's experiment. The team was using iron from a decommissioned battleship to filter out other particles. This involved packing the grooves inside one of the battleship's gun barrels with wire wool in order to produce a uniform volume of material. After an especially hard day crawling along the gun barrel, a rather upset graduate student stormed into Lederman's office to resign. 'You are never going to get me down that gun barrel again' he is supposed to have said. Lederman's reply is worthy of

a Nobel Prize in itself: 'You must go back, we have no other students of your calibre!'

[6] You may be wondering if we can really say that a force is involved as there is only one particle on the left-hand side—how can a force act on one particle? This is a very deep subject in particle physics we tackle in chapter 8.

[7] A remark attributed to the theorist Isidore Rabi on hearing of the existence of the muon and its similarity to the electron.

[8] Carl Anderson and Seth Neddermeyer first observed muon tracks in the early 1930s.

[9] Strictly, the reactions produce particles called pions which then decay into muons; see chapter 8.

[10] Lifetimes of this size are characteristic of the involvement of the weak force.

[11] This very odd sounding prediction has been confirmed by flying an atomic clock in a high-speed plane. Before the flight it was synchronized with a duplicate clock and when compared again after the flight was found to have lost time.

[12] This is quite a tricky argument involving conservation of energy and momentum. We will look at it again in more detail when we discuss neutron decay in chapter 8.

Chapter 5

Antimatter

In this chapter we shall examine the generation structure of weak force reactions. Lepton numbers are introduced as internal properties and extended to cater for antileptons. We shall anticipate chapter 6 by introducing baryon number and the antiquarks. Some general, but important, remarks will be made about the nature of antimatter. Finally, we shall consider lepton/antilepton reactions.

5.1 Internal properties

The weak interaction can change a lepton into its generation partner, but this conversion cannot take place between different generations. The weak force also acts predominantly within quark generations, but it is able to cross generations with a reduced effect. The conservation of leptons rule is based on this information. However, the rule seems to get into trouble when applied to the decay of particles such as:

$$\mu^- \rightarrow e^- + \nu_\mu + \nu_? \qquad (4.11)$$

$$\tau^- \rightarrow \mu^- + \nu_\tau + \nu_?. \qquad (4.12)$$

One of the produced neutrinos must be of the same generation as the original particle to conserve the number of particles in that generation. The second neutrino, however, is a complete mystery. The solution to this puzzle reveals what it is that makes the lepton generations so distinct.

The electron, muon and tau are very similar particles, the only difference between them being mass. On its own, this difference might be enough to separate them into distinct generations. However, the situation is far less clear for the three neutrinos. There is no compelling evidence, at the moment, that any of them have any mass at all, never mind different masses. If that is the case, then we are hard pressed to say what the physical difference between them is! They are all zero mass, zero charge objects. Yet Lederman's work convinced us of the difference between the ν_e and the ν_μ. There must be some property that distinguishes them, even if we cannot measure it.

Let us consider the possibility that there is some *internal* property that we cannot measure by any conventional means (such as we might use to measure charge and mass) which distinguishes the generations of leptons. We call this property *lepton number* and it splits into three different 'values' L_e, L_μ and L_τ. Table 5.1 shows how this new property is assigned to the various leptons.

Table 5.1 Assignments of lepton numbers.

	Electron-number L_e	Muon-number L_μ	Tau-number L_τ
electron	1	0	0
electron-neutrino	1	0	0
muon	0	1	0
muon-neutrino	0	1	0
tau	0	0	1
tau-neutrino	0	0	1

Imagine that there is a 'switch' that sits inside every lepton. This switch can be set in one of three ways that determines the lepton generation that the particle belongs to. To record the position of the switch we need three numbers L_e, L_μ and L_τ. These numbers can be either 1 or 0. A member of the first generation has $L_e = 1$ but $L_\mu = 0$ and $L_\tau = 0$. Of course we do not need to keep on saying which numbers are zero, it is sufficient to identify the number that is equal to 1.

Notice that the switch does not distinguish between the lepton and its neutrino: that is done by mass. Lepton number only distinguishes between generations. Experiments cannot read the value of this switch directly, but we can tell how it is set by observing the particle's reactions with other particles. Of course this switch does not actually exist—there are no mechanisms inside a particle—this is just a way of thinking about lepton number.

One reason why we cannot measure such a property is that it does not have a size, in the sense that charge or mass does. Electron-number cannot be 1.4 or 2.7 or 1.6×10^{-19} or any other number. A particle has either got electron-number or it hasn't—they are the only two possibilities. Physicists call such properties of particles (and we shall come across others) *internal properties*[1].

So far all we have done is to 'invent' a series of numbers that separate the various lepton generations. A physicist would need to be convinced that we are doing something more interesting than just playing with numbers. We need to ensure that lepton number is a real physical property.

There are two basic conditions that, if satisfied, will convince most physicists that a given property is real and not an invention of our imagination:

- it can be demonstrated that lepton number is a conserved quantity in particle reactions;
- using the lepton number conservation rule helps us to understand reactions that we could not understand or predict otherwise.

Happily, both of these conditions are satisfied.

5.1.1 Lepton number conservation

Consider a reaction from the previous chapter:

$$\nu_e + d \rightarrow u + e^-. \tag{4.3}$$

By definition all quarks have lepton number = 0 (they are not leptons!). The ν_e and the electron both have $L_e = 1$. Hence the total lepton numbers of the particles before and after the reaction are:

$$\nu_e \ + \ d \ \rightarrow \ u \ + \ e^-$$
$$L_e \quad 1 \ + \ 0 \ = \ 0 \ + \ 1.$$

Evidently L_e is conserved in this reaction. We need not consider L_μ and L_τ as the only leptons in the reaction are from the first generation.

Taking another example:

$$\nu_\mu + e^- \ \rightarrow \ \mu^- + \nu_e. \tag{4.8}$$

In this case the total electron-number (L_e) and the total muon-number (L_μ) *separately* must be the same before and after the reaction:

$$\nu_\mu \ + \ e^- \ \rightarrow \ \mu^- \ + \ \nu_e$$
$$L_e \quad 0 \ + \ 1 \ = \ 0 \ + \ 1$$
$$L_\mu \quad 1 \ + \ 0 \ = \ 1 \ + \ 0.$$

We could consider many other reactions that would all demonstrate the conservation of the various lepton numbers. This has become a well-established rule in particle physics amply confirmed by experiment and with a secure theoretical grounding as well.

LEPTON NUMBER CONSERVATION
The total electron-number, muon-number and tau-number are separately conserved in all reactions

---

Note the similarity between this rule and the conservation of electrical charge.

By this stage you may have the uncomfortable feeling that all we are doing is describing what is essentially a very simple thing in a complex way. In the last chapter we simply said 'the weak force must maintain the number of leptons of each generation that are involved in a reaction'. Now we are conserving three different lepton numbers.

This would be a serious criticism if it could not be shown that lepton number conservation leads us into new physics that we could not otherwise explain.

5.1.2 Mystery neutrinos

There are some reactions that seem to violate the rule that we now call lepton number conservation. One of the reactions that is giving us trouble is:

$$\mu^- \rightarrow e^- + \nu_\mu + \nu_?. \tag{4.11}$$

If we look at this reaction in terms of lepton number then something interesting emerges:

	μ^-	$\rightarrow$	e^-	$+$	ν_μ	$+$	$\nu_?$
L_μ	1	$=$	0	$+$	1	$+$	?
L_e	0	$=$	1	$+$	0	$+$	?

The assignment of lepton number for the mystery neutrino is unclear as we have not identified what sort of object this is. One thing seems very likely from this however—*it must have muon-number = 0*. If this were not the case, the muon-number conservation law would certainly be violated.

Notice also that initially the electron-number was zero, and then the decay created a particle with an electron-number of 1. The total on the right-hand side would be zero if our mystery neutrino had electron-number = −1.

Now this *looks* like a mathematical invention. Certainly if the '?' in the electron-number row were −1 then the totals would balance, but this alone does not make it physically correct. Let us for the moment suppose that our mystery neutrino has electron-number −1 and see if that helps to explain any other puzzles.

Another reaction that has been giving us trouble is the β decay of some nuclei:

$$n \rightarrow p + e^-. \tag{4.9}$$

This reaction is due to the more basic quark reaction inside the neutron:

$$d \rightarrow u + e^-. \tag{4.10}$$

When we discussed this reaction in the previous chapter I suggested that all was not quite as simple as it seemed. If we examine the reaction in terms of lepton number:

$$d \quad \rightarrow \quad u \quad + \quad e^-$$
$$L_e \quad 0 \quad \neq \quad 0 \quad + \quad 1.$$

then we see that L_e is not conserved. This dilemma can be solved by suggesting that our mystery neutrino is produced in the reaction as well:

$$d \quad \rightarrow \quad u \quad + \quad e^- \quad + \quad \nu_?$$
$$L_e \quad 0 \quad = \quad 0 \quad + \quad 1 \quad + \quad -1.$$

Experimentally it is easy to show that the mystery neutrino is produced in this reaction.

Historically it was in order to solve another puzzle connected with reaction (4.9) that Pauli first suggested the existence of the neutrino type of particle (see page 147).

Now we have made real progress. The presence of the mystery neutrino in (4.9) was deduced entirely through trying to conserve electron-number. This is concrete evidence that the property exists and is not just an invention of physicists. The next thing to do is see how this neutrino reacts with matter.

We are used to neutrinos reacting with the nuclei of atoms, so we can imagine an experiment in which the mystery neutrinos are allowed to pass through a block of matter in order to see what reactions take place. If this experiment were to be done[2] then something quite remarkable would be discovered. A totally new particle would be produced!

$$\nu_? + p \quad \rightarrow \quad n + e^+. \tag{5.1}$$

The e^+ has a positive electrical charge equal to that of the proton—but it is not a proton as the mass can be measured rather easily and found to be exactly the same as the mass of the electron.

This, as one might imagine, is an important discovery. Historically, the existence of this particle has been known for some time as it was discovered, in rather different circumstances to those described, in 1933[3]. It has been named the *positron*.

Within the context of our discussion, the significance of the positron is that it must have $L_e = -1$ (like the mystery neutrino). This follows from applying the conservation law:

$$
\begin{array}{ccccccc}
& \nu_? & + & p & \rightarrow & n & + & e^+ \\
L_e & -1 & + & 0 & = & 0 & + & -1.
\end{array}
$$

Notice that the proton and neutron have lepton numbers of 0: they are composed of quarks.

At this point we need to stop and consolidate. This has been a complex section introducing some new ideas and we need to ensure that they have all sunk in before we can go any further.

COFFEE POINT
stop reading
make a cup of coffee
sit and think over the following points

- We have introduced the idea of internal properties to describe the difference between particles that have no other obvious physical difference;
- the internal property that distinguishes the generations of leptons is called lepton number;
- the total lepton number is a conserved quantity in many reactions;
- some reactions do not obviously conserve lepton number;
- we can extend the idea of lepton number by suggesting that the mystery neutrino produced in muon decay has $L_e = -1$;
- this new neutrino also turns up in β decay where it was not expected and helps to solve the generation problem that seemed to occur with this reaction;
- if this new neutrino is passed through matter, then it can react with a proton to produce a positively charged particle with the same mass as the electron—this particle has been named the positron;
- positrons also have electron-number -1.

This is a convincing argument for the physical reality of internal properties.

5.2 Positrons and mystery neutrinos

It seems as if we have introduced a new generation of leptons. As well as the first generation (e^-, ν_e), in which both particles have $L_e = 1$, we have the positron and the mystery neutrino, both of which have $L_e = -1$. However, by definition, a new generation would have to have $L_e = 0$ and a new type of lepton number to identify it. What we have here are particles that are *related* to the first generation, but not standard members of it. They represent a sort of 'inversion' of the first generation.

It is sensible now to stop referring to the 'mystery neutrino': we may as well give it its full name: the *electron antineutrino*, symbolized $\bar{\nu}_e$. The bar over the top of the symbol is to show that it has $L_e = -1$ rather than $+1$ as in the case of the ν_e.

We can now write the full muon decay equation and the full β decay equation:

$$\mu^- \rightarrow e^- + \nu_\mu + \bar{\nu}_e \qquad (5.2)$$

$$n \rightarrow p + e^- + \bar{\nu}_e. \qquad (5.3)$$

We have not yet mentioned the tau decay. Is there also a neutrino with $L_\mu = -1$? A muon antineutrino? Yes, there is!

$$\tau^- \rightarrow \mu^- + \nu_\tau + \bar{\nu}_\mu. \qquad (5.4)$$

The $\bar{\nu}_\mu$ is produced in exactly the same manner as the $\bar{\nu}_e$ in the muon decay. A simple inspection of the muon- and tau-numbers will show that all is well and both are conserved.

Of course, we are now led to the next question: what happens when $\bar{\nu}_\mu$ passes through matter?

$$\bar{\nu}_\mu + p \rightarrow n + \mu^+. \qquad (5.5)$$

As one might expect, there is a new particle produced which is exactly the same mass as the muon, but with a positive electrical charge. It is called the *antimuon*.

The pattern, which is shown in table 5.2, is completed by the *antitau* and the *tau antineutrino*.

Table 5.2 The extended lepton families.

	1st generation	2nd generation	3rd generation
lepton number $= +1$	$\begin{bmatrix} e^- \\ \nu_e \end{bmatrix}$	$\begin{bmatrix} \mu^- \\ \nu_\mu \end{bmatrix}$	$\begin{bmatrix} \tau^- \\ \nu_\tau \end{bmatrix}$
lepton number $= -1$	$\begin{bmatrix} e^+ \\ \bar{\nu}_e \end{bmatrix}$	$\begin{bmatrix} \mu^+ \\ \bar{\nu}_\mu \end{bmatrix}$	$\begin{bmatrix} \tau^+ \\ \bar{\nu}_\tau \end{bmatrix}$

In this table 'lepton number $= -1$' should be taken to mean that either the electron-number or muon-number or tau-number $= -1$, depending on the generation.

By this time the reader may be feeling somewhat aggrieved. In chapter 1 I listed six material particles from which all matter in the universe is made, and now, four chapters further on, we are discussing six more objects that were not present in that list.

Technically speaking the list in chapter 1 was perfectly complete—it includes all the fundamental *matter* particles. These new objects, with lepton numbers $= -1$, are known as *antimatter* particles.

If you thought that antimatter was the creation of science fiction and only existed in the engine room of the Starship Enterprise, then I am afraid that you are wrong. Antimatter exists. It can be created in the lab and particle physicists[4] frequently use it in their experiments. Some of the stranger properties that writers tend to give antimatter (like it producing antigravity) are not true, however (antiparticles fall in a gravitational field, they do not rise!).

Antimatter particles have the same mass as the equivalent particle, opposite charge, *if the particle has charge*, and, at least in the case of the leptons, opposite lepton number.

Antimatter is remarkably rare in the universe. Nobody is totally sure why the universe does not consist of equal amounts of matter and antimatter. If there is a large amount of antimatter in the universe, then it is hidden somewhere[5].

In summary, we have six leptons split into three generations and six antileptons also split into three generations. The antileptons reflect the properties of the leptons, most obviously by a reversal of charge and lepton number.

5.3 Antiquarks

We arrived at the idea of antileptons through the introduction of the three lepton numbers that distinguish between the generations. Perhaps we can extend the idea to quarks by introducing three quark numbers to distinguish the quark generations?

We could define a 'down-number' that the d and u quarks have, a 'strange-number' that the s and c quarks have and a 'bottom-number'(!) possessed by b and t. Unfortunately, this does not work.

If you remember, the weak force does not totally separate the quark generations as it does the lepton generations. This would imply that the 'up-number', etc, would not be conserved quantities in the weak interaction, hence they would not pass one of the tests that we applied to lepton number to be sure that it was a real property. One could introduce these numbers, they just would not tell us anything useful about *all* weak reactions.

The situation is not totally lost. There is a property that distinguishes the quarks—that they are not leptons! The weak force does *absolutely* distinguish quarks from leptons; it cannot turn a quark into a lepton.

Another way of seeing this is to remember that all the quarks have all three lepton numbers $= 0$. This is equivalent to not having a lepton number 'switch'. Although quarks do not carry lepton number, they do carry a different form of 'internal switch'—one that makes them quarks rather than leptons. This internal property is referred to as *baryon number* (B). All quarks have baryon number $= 1/3$, all leptons have baryon number $= 0$. The weak force, as well as the other forces, conserves baryon number:

> ## CONSERVATION OF BARYON NUMBER
> In all reactions the total baryon number of the particles
> before the reaction must be the same as the total baryon number
> after the reaction.

If the leptons were not strictly divided into generations by the weak
force, then there would only be a single lepton number rather than
three, and the situation for the quarks and leptons would be identical.

Two things seem odd straight away. Why is the property called baryon
number not quark number, and why do quarks have baryon number
$= 1/3$ rather than 1?

Baryon number was invented before quarks were discovered. Its name
was coined independently from the name 'quark'. We shall see where
the name comes from in chapter 6. Quarks carry $B = 1/3$ because
baryon number was originally defined with the proton in mind. The
proton is assigned baryon number $= +1$. As protons contain three
quarks, each quark has baryon number $= 1/3$: the total baryon number
of the proton is the sum of the baryon numbers of the particles from
which it is composed.

This is a subtle point that needs some further discussion. Baryon
number, and indeed the lepton numbers, are supposed to represent
internal properties of particles. Such properties do not have a size
in the sense that electrical charge or mass have a size. This being the
case, *it does not matter what number we use to represent them*. There
would be no problem in saying that the electron has $L_e = -42.78$, as
long as we also said that the ν_e had $L_e = -42.78$ and that the e^+ had
$L_e = +42.78$. It would be a rather silly number to deal with, but the
size of the number is irrelevant. When we represent binary numbers in
electronics we use a similar convention. For example, binary 1 is taken
as $+5$ volts and binary 0 as 0 volts. In principle any pattern of voltage
would have done the job.

The pattern we have chosen for baryon number is based on calling
the proton the $B = +1$ particle. The only possibilities for the baryon
numbers of *fundamental* particles are $B = +1/3, 0, -1/3$. $B = +1/3$
are the quarks, $B = 0$ are the leptons and $B = -1/3$ are the *antiquarks*.

Just as for the leptons, each quark has an antiquark partner. Each antiquark has the opposite charge but the same mass as its partner matter quark.

Putting all this into a table like that for the leptons we obtain table 5.3.

Table 5.3 The extended quark families.

	1st generation	2nd generation	3rd generation
baryon number $= +\frac{1}{3}$	$\begin{bmatrix} u \\ d \end{bmatrix}$	$\begin{bmatrix} c \\ s \end{bmatrix}$	$\begin{bmatrix} t \\ b \end{bmatrix}$
baryon number $= -\frac{1}{3}$	$\begin{bmatrix} \bar{u} \\ \bar{d} \end{bmatrix}$	$\begin{bmatrix} \bar{c} \\ \bar{s} \end{bmatrix}$	$\begin{bmatrix} \bar{t} \\ \bar{b} \end{bmatrix}$

The antiquarks are symbolized, as in table 5.3, by the equivalent quark symbol with a '-' or bar over the top. An antiup quark is written $\bar{u}$ and pronounced 'u bar'.

The strong interaction acts between antiquarks just as it does between quarks, implying that antiquarks combine to form antiparticles. The most obvious examples of this are the *antiproton*, a $\bar{u}\bar{u}\bar{d}$ combination, and the *antineutron* $\bar{n}$, $\bar{u}\bar{d}\bar{d}$.

The antiproton will have the same mass as the proton, but a negative electrical charge. The antineutron, however, will still have a zero charge, but is different from the neutron (despite having the same charge and mass) because it is composed of antiquarks rather than quarks.

We shall return to the subject of combinations of antiquarks when we have discussed in more detail the combinations of quarks—something that we will do in chapter 6.

5.4   The general nature of antimatter

So far we have introduced antileptons as particles with negative lepton numbers and antiquarks as having negative baryon numbers. The justification for doing this is that it helps us make rules about lepton and baryon number conservation. Of course, this would just be playing with numbers if it were not for the fact that these antiparticles exist and can be created in the lab. However, there is far more to antimatter than this simple introduction would suggest.

The first hint of the existence of antimatter came, in 1928, as a consequence of the work of the English theoretical physicist P A M Dirac. Dirac had been doing research into the equations that govern the motion of electrons in electric and magnetic fields. At slow speeds the physics was well understood, but the discovery of relativity in 1905 suggested that the theory was bound to go wrong if the electrons were moving at speeds close to the speed of light.

Dirac was the first person to guess the mathematical equation that correctly predicted the motion in these circumstances[6]. The Dirac equation, as it is now known, has become a vital cornerstone of all theoretical research in particle physics. Unfortunately, the mathematics that we would need to learn to be able to deal with this equation would take far too long to explain in a book like this.

When Dirac came to solve his equation he discovered that it produced two solutions (in the same way that an equation like $ax^2 + bx + c = 0$ will produce two solutions). One solution described the electron perfectly, the other a particle with the same mass but opposite electrical charge. This was a considerable worry to Dirac as no such particle was known, or suspected, at the time.

No matter what mathematical tricks Dirac tried to pull, he could not get rid of this other solution. At first he thought that the second solution must represent the proton, but the huge difference in mass made this untenable. The problem was resolved when Anderson discovered the positron in 1933. This particle was the second solution to Dirac's equation.

Dirac unknowingly predicted the existence of antimatter. Since that time many famous physicists have worked on the general theory of antimatter[7].

We now know that the existence of antimatter follows from some very basic assumptions about the nature of space and time. The existence of both matter and antimatter is necessary, given the universe in which we exist. This is a very deep and beautiful part of particle physics.

Any theory that is consistent with relativity must contain both particles and antiparticles.

Unfortunately I can think of no adequate way of explaining this at the level that this book intends. I must content myself with saying that the very basis on which the universe is set up requires that every fundamental particle must have an antiparticle partner[8].

5.5   Annihilation reactions

Science fiction writers have used antimatter for many years, so it is hardly surprising that some people's knowledge is influenced by such stories. Unfortunately, the writer does not always stick to the truth when it suits the story better to bend the facts. However, one fact is accurately portrayed: matter and antimatter will destroy each other if they come into contact.

No-one has ever mixed large amounts of antimatter and matter together to produce an explosion of atomic bomb proportions. This is because antimatter is difficult to make in anything other than minute amounts[9].

However, particle physicists regularly collide small amounts of matter and antimatter together because of the energy and new particles that can be produced in such reactions. These are called *annihilation reactions*. A typical example of an annihilation reaction is:

$$e^+ + e^- \rightarrow ?$$

We shall come to understand how important this reaction has been for the development of particle physics.

Let us, for the moment, not worry about the practical details of how to arrange for such a reaction to take place and how to detect the particles that are produced. Instead we need to examine the fundamental features that emerge. Imagine that we have two beams, one of e^+ and the other

of e⁻, that we can allow to come together at will with any energy we
choose. Examining the reactions that take place when the beams collide
reveals several interesting features.

- At all energies the e^+ and the e^- can simply bounce off each other
  without anything interesting happening.
- If the e^+ and the e^- are moving very slowly, then they can go
  into orbit round each other producing an atom-like object called
  positronium. Positronium is very unstable: the e^+ and the e^- will
  invariably destroy each other.
- If we increase the reaction energy, i.e. the kinetic energy of the
  e^+ and the e^- entering the reaction, then something interesting
  happens. Once we have exceeded a precise threshold energy, the
  following reaction starts to take place:

$$e^+ + e^- \rightarrow \mu^+ + \mu^-. \tag{5.6}$$

  Inspection of the lepton numbers soon indicates that all is well;
  the reaction does not violate any conservation law. However, it is
  quite a surprising thing to see happening.
- If the reaction energy is turned up some more, then both reactions
  continue to happen until another threshold is passed, at which point
  another reaction[10]:

$$e^+ + e^- \rightarrow \tau^+ + \tau^- \tag{5.7}$$

  starts to take place.
- In between these thresholds some other boundaries are passed as
  well. As the energy is increased particles are produced indicative
  of the presence of quarks. These 'jet' reactions, as they are called,
  are not as 'clean' as the lepton reactions. The strong interaction
  makes them complicated. For the moment we shall not study
  them. They shall form an important section of chapter 9, as they
  represent direct evidence for the existence of quarks.

When the e^+ and the e^- react together the initial combination of
particles is such that the electrical charge, lepton number and baryon
number all total to zero. This allows the reaction to produce anything,
provided the combination adds to zero again. Any generation of lepton,
or any generation of quark, can be produced provided the appropriate
antiparticle is produced as well. Hence, annihilation reactions are an
easy way of searching for new particles.

The materialization of particles and antiparticles is a consequence of the relationship between mass and energy contained in the theory of relativity. There must be enough energy in the reaction to produce the pair of particles. For example, the total energy of the e^+ and e^- needed to produce a tau pair is approximately 16.96 times that required to produce a muon pair as the mass of the tau is 16.96 times the mass of a muon.

This is such an important point that it needs summarizing:

> Annihilation reactions between matter and antimatter can lead to the production of new matter and antimatter provided they are produced in equal amounts and provided there is enough energy in the reaction.

This annihilation of matter and antimatter, leading to the creation of new matter and antimatter, seems to imply an essential symmetry between them. Every time matter and antimatter annihilate each other some new combination of matter and antimatter is produced, so it is difficult to see how there could be different amounts of matter to antimatter in the universe. If there are equal quantities of matter and antimatter, then where is all the antimatter? This returns us to one of the fundamental problems in cosmology—the creation of the universe would violate our conservation laws, unless equal amounts of matter and antimatter are produced!

5.6 Summary of chapter 5

- There are three types of lepton number which are conserved in all reactions;
- each type of lepton number is characteristic of one of the generations;
- the lepton number represents a new, internal, property of leptons;
- particles with negative lepton numbers exist and are called antileptons;
- for each lepton there is an equivalent antilepton;

- a charged lepton and its antilepton partner have the same mass but opposite electrical charge and lepton number;
- neutral leptons and antileptons differ in their lepton numbers;
- quarks have a property called baryon number that is conserved in all reactions;
- baryon number is not generation-specific;
- every quark has an antiquark partner with opposite charge and baryon number;
- matter/antimatter annihilation reactions lead to new forms of matter being produced;
- these annihilation reactions point to a relationship between matter and energy.

Notes

[1] There are various terms used for internal properties: they are sometimes referred to as intrinsic properties, or quantum numbers.

[2] Indeed, this experiment can be done, it is just that the neutrinos are produced in a different way.

[3] Its discoverer, Anderson, won the Nobel Prize for finding the positron in a cosmic ray experiment.

[4] Even the ones without pointed ears!

[5] It is difficult to see how large amounts of antimatter could be hidden. Presumably, it would have to be in some separate region of the universe. However, at the boundary between the matter and antimatter regions reactions would take place that would produce enormous quantities of gamma rays. These would be easily observed by astronomers, but have not been seen.

[6] Dirac was awarded the Nobel Prize in 1933.

[7] This is why the positron is not called the antielectron. When it was discovered, physicists had no idea of the general nature of antimatter.

[8] This is not to say that both must exist at the same time. Theory says that each particle *type* must also have an antiparticle *type*, not that every particle in the universe at this moment must have an antiparticle partner somewhere.

[9] In September 1995 a team lead by Professor W Oelert working at CERN managed to produce atoms of antihydrogen—antiprotons with antielectrons in orbit. The experiment used the low energy antiproton ring (LEAR) at CERN. During its three week run, the experiment detected nine antiatoms. See Interlude 2 for more information.

[10] This is how the tau was discovered!

Chapter 6

Hadrons

In this chapter we shall study quarks and their properties in more detail. This is a more complicated task than that for the leptons. All quarks and antiquarks are bound into composite particles by the strong interaction. In order to study the quarks, we must study these particles. This chapter is the first of three closely related chapters in which we shall study the properties, reactions and decays of hadrons.

6.1 The properties of the quarks

It is not possible to present a table of the physical properties of the quarks with the same degree of confidence as we did for the leptons in chapter 4. The electrical charges of quarks can be stated with some certainty, but the masses of the quarks are a source of controversy.

6.1.1 Quark masses

Many textbooks quote values of quark masses that are similar to those in table 6.1.

All such tables should be taken with a pinch of salt. Other textbooks quote quark masses that are very different from the ones I have chosen to use. The values of masses that are used depend on the calculations that are being carried out. The values of mass that I have presented in

Table 6.1 The masses of the quarks (in GeV/c^2).

Charge	1st generation	2nd generation	3rd generation
$+2/3$	up 0.33	charm 1.58	top 180[1]
$-1/3$	down ~ 0.33	strange 0.47	bottom 4.58

table 6.1 are those that are most appropriate for the use to which we will put them.

The key issue here is the nature of the strong force. The strong force binds quarks together into composite particles. We believe that it is theoretically impossible to isolate an individual quark (or antiquark). This makes it impossible to directly measure the mass of an individual quark. Quark masses have to be deduced from the masses of the composite particles that they form. However, this is not as simple a task as it sounds.

The strong force holds the quarks together tightly, so very large energies are involved inside the particles. As we know from chapter 2, energy implies mass so some of the mass that we measure for a composite particle is due to the masses of the quarks, and some is due to the energy of the forces between the quarks. This makes deducing the masses of the individual quarks very difficult. Imagine that you were trying to estimate the mass of an ordinary house brick, and all the information that you had was the masses of various houses. Even if you knew exactly how many bricks there were in each house, the bricks are of different types and you need to take into account all the other materials that go to make up the house as well. Such is the task of the physicist trying to deduce quark masses.

At the moment it is impossible to calculate how much of the mass of a proton, say, is due to the masses of the quarks inside and how much is due to the interaction energies between the quarks. On the other hand, when these composite particles interact with one another the result is often due to the interactions of the individual quarks within the particles, hence the masses of the quarks have some influence on the reaction. This gives us another way of calculating the quark masses.

Unfortunately, these two techniques result in very different values for the quark masses. At the moment we do not have a complete mathematical understanding of the theory, so the puzzle cannot be resolved.

The values that are quoted in table 6.1 are values deduced from the masses of composite particles. Given the degree of uncertainly associated with table 6.1, what can we reasonably deduce from it? The most important point to gain from this table is the *relative values* of the quark masses. The antiquark masses are assumed to be the same as the quark masses: the charges are, of course, reversed.

The similarity in the masses of the u and d quarks is well established experimentally. After all the proton (uud) and the neutron (udd) are very nearly the same mass. Other particles composed of u and d combinations also have small mass differences, which emphasizes how close the u and d are. The d quark is slightly more massive than the u— if this were not the case neutrons (udd) would not be unstable particles that decay into protons (uud).

Charm, strange and bottom are now quite familiar quarks although our experiments have only recently been of high enough energy to produce enough b quarks for them to be systematically studied.

The chronology of discovery is interesting. The c quark was a surprise discovery in 1974 and the b came soon after in 1977. However, despite great effort by experimenters we had to wait until 1994/95 for the top quark to be discovered. This is a reflection of the much greater mass of this quark. Experiments had to reach much higher energies to produce particles that contained t quarks.

6.1.2   Internal properties of quarks

We came across one internal property of quarks in chapter 5—baryon number. All quarks have baryon number $= 1/3$ and all antiquarks have baryon number $= -1/3$. Strictly speaking, baryon number is the only internal property of quarks. They do not have quark numbers like the lepton numbers, because the weak force does not respect the quark generations.

Physicists have found it useful to label quarks with 'properties' that do not pass all our tests for internal properties (they are not conserved by the weak force), but work quite well for the other forces. These are useful as they help us to keep track of what might happen when quarks interact. These properties are not referred to as quark numbers but as 'flavour numbers'.

Flavour is another whimsical term that is commonly used in the particle physics community. Particle physicists like to refer to the quarks as having different flavours—up, down, strange, etc. There are six flavours of quarks, and six flavours of leptons. Table 6.2 lists the assignment of flavour numbers. Note that the flavour numbers belong to *each quark*, not to *each generation* as in the case of the leptons. Electron number is a property of the e$^-$ *and* the ν_e. There is no common flavour number to the u and d quarks. Each quark has its own distinctive flavour number.

Table 6.2 The quark flavour numbers.

Quark	U	D	C	S	T	B
up	1	0	0	0	0	0
down	0	−1	0	0	0	0
charm	0	0	1	0	0	0
strange	0	0	0	−1	0	0
top	0	0	0	0	1	0
bottom	0	0	0	0	0	−1

The antiquarks also have flavour numbers, the values being the negative of the corresponding particle flavour number, i.e. $\bar{c}$ has $C = -1$. The symbols S and C refer to 'strangeness' and 'charm' (not strangeness-number: although this name was used, it rapidly went out of fashion). T and B are 'topness' and 'bottomness'. There was a strong lobby for 'truth' and 'beauty' by the romantic element of the particle physics community, but the names did not gain much support.

U and *D* should refer to 'upness' and 'downness', but as these properties are hardly ever used they have not gained familiar names. It has never proved useful to use *U* and *D* numbers as the up and down quarks are so similar in mass. *U* and *D* have been included for completeness only.

These flavour numbers are very useful for keeping track of what happens to the various quarks when particles react together. We shall use them extensively in this way in the next chapter.

It may seem odd that some of the flavour numbers are -1. For example the strange quark has $S = -1$ (which means that the antistrange quark has $S = +1$). This is a consequence of the properties being defined *before* the discovery of quarks, as baryon number was. Unfortunately, the particle that was first assigned strangeness $+1$ contained the $\bar{s}$ quark, not the s. Hence we are stuck with the s quark having $S = -1$.

Before we can use the flavour numbers to study the reactions of composite particles, we must understand the basic features of the strong force that binds the quarks into these particles.

6.2   A review of the strong force

The strong force only exists between quarks. The leptons do not feel the strong force.

The strength of the force varies with distance, much more so than gravity or the electromagnetic force. The strong force effectively disappears once quarks are separated by more than 10^{-15} m (about the diameter of a proton). For distances smaller than this, the force is very strong indeed. The exact details of how the force's strength changes with distance are irrelevant at this stage and so we will postpone them until chapter 7.

The strength of the force does not depend on the type of quark that is involved, nor on the charge of the quark. Physicists say that the strong force is 'flavour independent'.

For the moment, it is best to think of the strong force as being purely attractive and so strong that it binds the quarks together in such a way as they can never escape. However, the properties of the strong force are such that the quarks don't *all* stick together in one large mass (otherwise the universe would be a huge lump of quarks). The strong force ensures that quarks and antiquarks can only stick together in groups of three (qqq or $\bar{q}\bar{q}\bar{q}$), or as a quark and an antiquark pair ($q\bar{q}$).

6.3 Baryons and mesons

The family name for composite particles made from quarks is *hadrons*. Before quarks were discovered, particle physicists thought that leptons and hadrons were the only constituents of the universe.

The hadron family tree has three branches—those that consist of three quarks bound together, those that are three antiquarks and those that consist of a quark and an antiquark.

A particle that is composed of three quarks is a member of the sub-group called *baryons*. If it is composed of three antiquarks, then it is an *antibaryon*. The quark–antiquark pairs are called *mesons*. The proton and neutron (uud and udd) are baryons, the π^+—pronounced 'pi plus'—(u$\bar{\text{d}}$) is a meson. Figure 6.1 shows the various categories of the hadrons.

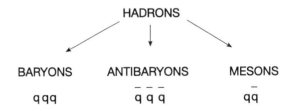

Figure 6.1 The hadron family tree.

Most of what we have to say about the baryons is also true about the antibaryons, except for obvious things like an antibaryon having the opposite electrical charge to the similar baryon, so we shall not refer to them explicitly unless we need to point out a different feature.

All baryons, except the proton, are unstable and will decay. The particles produced must include another baryon. Consequently, decay chains are formed as one baryon decays into another that is also unstable, etc. Such decay sequences must end at the proton, which is the lightest baryon and therefore stable.

6.4   Baryon families

If we restrict ourselves for the moment to just the lightest three quarks, u, d and s, then the number of possible ways in which they can be combined is a small enough number to be manageable. When the combinations are listed the order of the quarks is not important—uds would be the same as sud or dus. (By the way, we do not try to pronounce these as words, 'sud—a particle often met in washing machines', the individual letters are pronounced.)

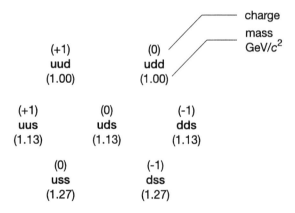

Figure 6.2   Quark combinations (predicted particle masses in GeV/c^2).

Figure 6.2 shows a selection of the possible combinations with their predicted masses in GeV/c^2. For the moment I have missed out uuu, ddd and sss. It is quite easy to use the properties of the quarks to deduce what baryons made from these combinations should be like. For example, the uus combination should have an electrical charge $= (+2/3) + (+2/3) + (-1/3) = 1$. Its mass should be $= (0.33 + 0.33 + 0.47)$ GeV/$c^2 = 1.13$ GeV/c^2. These properties have already been listed in figure 6.2, the charge above the combination and the mass below.

As one might expect, there is a gradual increase in the mass as the number of strange quarks increases. When we examine the actual particles that are formed we might also expect a small increase in mass along horizontal rows of the figure as the d quark is slightly more massive than the u.

The horizontal rows form *baryon families*. A family is defined as a set of baryons of similar mass with the same internal properties. The top row forms a family as the two particles have the same mass and the same strangeness ($S = 0$). The second row is a family of similar mass all with $S = -1$ and the bottom row the $S = -2$ family. This pattern of particle properties was noticed before the quark structure of baryons was known about. Indeed it was one of the important pieces of evidence that led Murray Gell-Mann and George Zweig to the discovery of quarks.

Figure 6.3 is a chart of various particles that have been discovered in reactions. Such a chart is sometimes called a *weight diagram* and they can be plotted in various ways. In this case I have grouped the baryons into family sets with mass increasing down the diagram. The diagram clearly follows the pattern of the quark combinations very well.

			strangeness	name
p		n	$S=0$	nucleon
(0.938)		(0.940)		
Σ^+	Σ^0	Σ^-	$S=-1$	sigma
(1.189)	(1.192)	(1.197)		
Ξ^0		Ξ^-	$S=-2$	cascade
(1.314)		(1.321)		

Figure 6.3 A baryon weight diagram (particle masses in GeV/c^2).

The mass predictions have not turned out to be exactly right, but this is to be expected given the uncertainly in the masses of the quarks and the nature of the forces between them. However, the pattern and its correspondence to that of the quark combinations are compelling.

There is one known baryon that is missing from this weight diagram. The lambda baryon (Λ) is known to be strangeness -1 (this can be deduced from the way it is produced) and has a mass of 1.115 GeV/c^2. This makes it a strong candidate to be a uds combination—the mass is about right and the strangeness value points to the presence of only one s quark. However, the uds combination would seem to be already taken by the Σ^0. In fact there is no reason why both particles should not have the same quark content.

One of the problems associated with calculating the masses of baryons is that the quarks inside have a variety of energy levels that they can occupy. This is a similar situation to that found in atoms in which the electrons in orbit round the nucleus have a variety of energy levels (see section 3.8). In an atom there are restrictions that must apply—no more than two electrons can be in the same energy level at any one time for example. Similar restrictions apply to quarks in baryons (and mesons) except the situation is more complicated as the energy levels are mostly determined by the strong forces between the quarks. In the case of the baryons and mesons the energy levels of the quarks have a significant influence on the mass of the particle—electronic energy levels hardly contribute to the mass of an atom.

Given a quark combination like uds there is greater freedom in how the quarks can be organized within the energy levels. The Σ^0 and the Λ masses reflect this—they have the same quark content but different masses as the quarks are in different energy levels. The Σ^+ and Σ^- have two identical quarks (Σ^+ uus, Σ^- dds) which restricts their energy levels more; they do not have a duplicate lower mass version like the Λ.

The complete weight diagram including the Λ is known as the *baryon octet*. This consists of quark combinations in low mass states. There is a similar set of particles in which the quarks occupy higher energy level states. Their properties are essentially the same as their lower mass brothers, except that their greater intrinsic energies make them more prone to decay.

The uuu, ddd and sss combinations can *only* exist in these higher mass states. Having three identical quarks prevents them from occupying energy levels in such a way as to produce low mass versions. This is why I omitted these combinations in figure 6.2. In figure 6.4 the set of higher mass particles is grouped into a weight diagram. Notice that the delta family contains two particles which have exactly the same quark combinations as protons and neutrons—the Δ^+ is a uud, like the proton and the Δ^0 is a udd, like the neutron. Essentially these are higher mass versions of the proton and neutron. However, the delta family goes on to include the Δ^{++} which is uuu and the Δ^- which is ddd. Neither of these combinations can be found in the lower mass table.

				strangeness	name
Δ^- ddd (1.23)	Δ^0 udd (1.23)	Δ^+ uud (1.23)	Δ^{++} uuu (1.23)	S=0	delta
	Σ^{*+} dds (1.383)	Σ^{*0} uds (1.384)	Σ^{*-} uus (1.387)	S=-1	sigma*
		Ξ^{*0} uss (1.532)	Ξ^{*-} dss (1.535)	S=-2	cascade*
		Ω^- sss (1.67)		S=-3	omega

Figure 6.4 The higher mass baryons.

This diagram is known as the *baryon decuplet*. Organizing the baryons and tabulating their properties in this way was named 'the eightfold way' by Murray Gell-Mann (who else?) after the Zen Buddhist path to enlightenment. Using his analysis, which eventually led him to suspect the existence of quarks, Gell-Mann *predicted* the existence of the Ω^- (sss) particle, which was discovered in 1964 with exactly the properties and mass that Gell-Mann suggested[2]. The sss does not occur in the octet diagram either so the Ω^- has no lower mass brother.

6.4.1 Higher mass baryons

Having spent some time discussing different baryons, it is rather sobering to realize that we have only employed half of the quarks that are at nature's disposal in the creation of particles. However, we have covered all the basic physics.

Simply adding the charm quark to our game adds enormously to the complexity. For example, the baryon decuplet becomes a three-dimensional pyramid. The base of the pyramid is formed from the decuplet, and the point at the top of the pyramid is the ccc combination. The full pyramid of combinations is illustrated in figure 6.5.

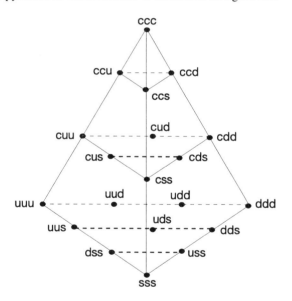

Figure 6.5 A charming extension to the baryon decuplet.

Of course, a similar extension must be made to the baryon octet as well (without the ccc combination).

At this point we start to run into the limits of our experimental knowledge. Baryons that contain the heavy quarks require a great deal of energy to produce—more than can be provided in many experiments. The Λ_c^+, which is udc and the Σ_c^{++}, which is uuc, were discovered in

1975 and the Λ_b (udb) in 1981. The study of these heavy flavour baryons is still being advanced.

6.5 Meson families

A meson is a hadron that consists of a quark–antiquark pair. The quark and the antiquark do not have to be of the same flavour. Any combination is possible, although some are less stable than others.

The u, d and s quarks (with their antiquarks) are the least massive quarks, and hence the most common in nature, so we might reasonably expect that mesons formed from them would also be quite common.

In figure 6.6 strangeness decreases as one moves down the diagram, but, in contrast to the baryon diagrams, mass does not increase. As expected there is a set of mesons that corresponds to these combinations (figure 6.7).

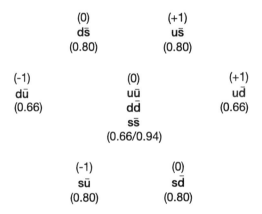

Figure 6.6 Meson quark combinations.

This time, as there are nine of them, the set is called the *meson nonet*.

The central row of three particles, the π^+, π^- and π^0 make up the pion family, the lightest mass and most common of the mesons. Occupying the same slot as the π^0 are two other particles the η and the η'.

		strangeness	family name
K⁰ (0.498)	K⁺ (0.494)	S=+1	kaons
π⁻ (0.140)　π⁰/η/η′ (0.135/0.547/0.958)　π⁺ (0.140)		S=0	pions/eta
K⁻ (0.494)	K̄⁰ (0.498)	S=-1	kaons

Figure 6.7 The meson nonet.

The π^0 is the u$\bar{\text{u}}$ *and* the d$\bar{\text{d}}$ combination. This does *not* mean that there are two types of π^0, one containing u$\bar{\text{u}}$ and the other containing d$\bar{\text{d}}$. As explained in chapter 8, the u$\bar{\text{u}}$ combination can change into the d$\bar{\text{d}}$ combination, and back again, inside the particle.

The η and the η' are different combinations of u$\bar{\text{u}}$, d$\bar{\text{d}}$ and s$\bar{\text{s}}$ states.

The kaon family is split on this diagram as the four particles have different strangeness. The K⁺ and the K⁻ are antiparticles of each other. One of the advantages of this diagram is that both types of particle are displayed on one picture.

Note that unlike the π^0, the K⁰ and K̄⁰ are different particles with slightly different properties. This seems quite strange (no pun intended). Both mesons have a zero electrical charge and the same mass, so what is the difference between them? The differences lie in the reactions that the two neutral kaons can take part in and the ways in which they decay. The situation is not so very different from that of the neutron. Its antiparticle has the same mass and the same zero electrical charge but, is a distinct object that reacts in different ways.

The situation is even more odd in the case of the neutrinos; their antiparticles are only different from the particle by having a different value of the lepton number and that is not something that can be measured directly—see chapter 5.

The masses of the particles have turned out to be rather less than one might expect given the quark masses from the previous section. This is typical of the problems associated with giving quarks mass. Values that are chosen do not seem consistent with all particles. This problem arises from the different ways in which the strong force binds the quarks together in a baryon (qqq) and a meson (q$\bar{\text{q}}$).

Other, less common, mesons containing heavier quarks have been created and studied in particle physics experiments. They are difficult to create, because of their large masses and their study, although much better developed than the study of heavy baryons, is still continuing.

6.6 Internal properties of particles

Particle physicists have chosen to label various hadrons by the internal properties of the quarks that they contain. To give an immediate example, any hadron that contains a strange quark is called a strange particle and given a value of the internal property S.

Strictly, S is a property that can only belong to a quark, but this convention was developed *before* quarks were discovered so, historically, physicists are more used to giving hadrons S values than they are quarks[3].

The K$^+$ was the particle chosen to define the values of S for all the others. By convention the K$^+$ is given $S = +1$. Notice that the quark combination of the K$^+$ is u$\bar{\text{s}}$—this is why it is the $\bar{\text{s}}$ that has $S = +1$, not the s quark, as might seem more sensible.

Despite its apparent eccentricity (in hindsight) the labelling of particles as if they had internal properties is very useful. The S value of the particle tells us how many strange quarks it contains (OK, it counts them in -1's, but nothing is perfect!).

For example, the Ξ^- has $S = -2$, so it must contain two strange quarks. As S values are quoted in particle data tables, we can work out what the quark content of a particle is by looking up a couple of numbers in a table.

To give a full example, let us work out the quark content of the Σ^+. From a data table (e.g. the back of this book):

(1) the Σ^+ is a baryon;
(2) the Σ^+ is a strange particle, with $S = -1$;
(3) the Σ^+ has a charge of $+1$.

The first point tells us that the Σ^+ contains three quarks (as that is the definition of a baryon). The second point implies that it contains a strange quark. In addition, the fact that it does not have any other internal properties (i.e. C, T or B) implies that the other quarks must be u or d. Noting from the table that the s quark has charge $-1/3$, we can work out that the other quarks in the Σ^+ must have a total charge of $+4/3$ in order to make the particle $+1$. The only way to make this from two quarks is uu. Hence the Σ^+ is an suu combination.

Other internal properties exist as well. The ones that are listed are C, B and T which cover particles that contain the charm, bottom and top quarks. For example the D^0 meson is a c$\bar{u}$ combination and so has $C = +1$. These internal properties count the respective quarks just as well as S does.

6.7 Summary of chapter 6

- It is difficult to give the quarks unambiguous masses;
- quarks have flavour numbers which are not really internal properties but which are quite useful 'book keeping' devices;
- particles composed of quarks are called hadrons;
- a meson is a q$\bar{q}$ combination;
- a baryon (antibaryon) is a qqq ($\bar{q}\bar{q}\bar{q}$) combination;
- there is a sequence of baryon families based on the proton and neutron;
- this sequence forms the baryon octet with increasing numbers of strange quarks;
- there is a sequence of baryon families (the decuplet) based on the Δ particles which have greater masses than the octet families;
- there are more particles in the decuplet than the octet as the symmetrical combinations (uuu, ddd, sss) are allowed in the decuplet but not in the octet;

- particles of different masses can have the same quark content as the quarks can be in different energy levels (which gives rise to the different mass);
- the symmetrical combinations can only occur if some of the quarks are in higher energy levels;
- there is a meson nonet which has a similar structure to the baryon octet;
- hadrons are tabulated in particle tables as having values of internal properties such as S, C, B and T;
- these are actually the flavour numbers of some of the quarks they contain;
- these internal properties of particles are useful clues to the quarks that they contain.

Notes

[1] This value is obtained from the 1994/95 Fermilab announcement of candidate top events. Notice that this fundamental particle is more massive than many decent sized molecules!

[2] The Ω^- prediction was made at a conference in CERN in 1962 by Gell-Mann and, independently, by Yuval Ne'eman. Unfortunately for Ne'eman, Gell-Mann was called to the blackboard before him.

[3] In a way this is a very good use of the term internal properties as the properties belong to objects inside the particle. Really and truly, they should be called flavour numbers as they are no more internal properties than the flavour numbers of quarks are. The only genuine internal properties we have come across are the lepton numbers and baryon number.

Chapter 7

Hadron reactions

In this chapter we shall use what we have learnt about their quark composition to discuss how hadrons react with each other. In the process we shall learn about various rules that help us to decide which reactions are physically possible. These rules, called conservation laws, are intimately related to the way in which quarks and leptons are acted upon by the fundamental forces.

7.1 Basic ideas

Particle reactions take place when two or more particles come close enough to each other for fundamental forces to cause an interaction between them. As a result of this interaction two things may happen:

(1) the particle's trajectory can change: they can be attracted, or repelled;

(2) the particles can change into different particles: it may be that a greater number of particles emerge from the reaction than entered it.

Of course, these possibilities are not mutually exclusive: if more particles are created, then the trajectories must have altered.

The fundamental forces at work and the energy of the reacting particles control the exact outcome of the reaction. As the energy increases a wider range of possibilities becomes open to the forces. Observing how the various possible reactions behave at different energies is an important technique in studying the fundamental forces.

The study of hadron reactions was the focus of particle physics in the 1950s and 1960s. As technology improved, experimenters collided particles at higher and higher energies. Physicists soon realized that the intrinsic and kinetic energies of the incoming particles were being mixed and redistributed by the fundamental forces. From this mixture new particles were materializing.

7.2 Basic processes

When two hadrons react with each other to form new particles it is invariably due to the strong force as:

- the gravitational force is far to weak to have any effect on fundamental particles;
- the weak force is much shorter range than the strong force, so a strong force reaction will tend to take place before the hadrons get close enough together for the weak force to come into play;
- the electromagnetic force is much longer range than either the weak or strong forces so hadrons coming together may well attract or repel each other via the electromagnetic force but at a distance; if they have enough energy to approach each other closely they will react via the strong force.

Early experiments collided protons with protons. This is quite easy to arrange experimentally. One technique is to use a tank of liquid hydrogen (called a bubble chamber) and to fire a beam of protons into the tank. Liquid hydrogen consists of protons and orbiting electrons. The electrons very rarely react with the protons passing through the liquid, so the tank is basically a collection of protons that will react with those in the beam, if they get close enough.

The most obvious reaction between two protons is:

$$p + p \rightarrow p + p \tag{7.1}$$

in which they simply bounce off each other (elastically scatter). This reaction can take place at any energy and is a consequence either of the electrostatic repulsion between the two positively charged protons, or of a strong force reaction which has not broken up the particles. As the energy in the reaction is increased (this corresponds to increasing the speed of the incoming protons) other strong force reactions start to take place such as:

$$p + p \rightarrow p + p + \pi^0. \tag{7.2}$$

Considering reaction (7.2) in more detail, one thing becomes apparent immediately: the total number of quarks and antiquarks in the reaction has changed:

p	+	p	→	p	+	p	+	π^0
u		u		u		u		u
u		u		u		u		$\bar{u}$
d		d		d		d		

The total number of d quarks has remained the same, but we have increased the number of u quarks by one. We have also picked up a $\bar{u}$.

This sort of process in which a quark and an antiquark have been materialized is the opposite to the annihilation reactions that we have seen before. Previously we have noted that an e^+e^- pair can annihilate into energy. Now we have the materialization of a $q\bar{q}$ from energy. This process is evidently not completely arbitrary, there are physical laws that must be followed. These laws place limitations on what the reaction is able to do and they are best expressed in terms of the conservation of certain quantities. In turn, these conservation laws tell us about the fundamental forces that are at work—in this case the strong force.

7.2.1 Conservation of charge

All materialization and annihilation reactions must involve matter and antimatter. Conservation of electrical charge prevents the strong force from materializing quarks such as u and $\bar{d}$ from energy (the total charge is not zero). However, a $u\bar{c}$ materialization would conserve charge. Yet, such a materialization is not seen in strong force reactions. The properties of the strong force dictate that the quark and antiquarks involved must be of the same flavour. You can't materialize a u and a $\bar{c}$ from the strong force, but you can materialize a u and a $\bar{u}$.

We have seen charge conservation at work in lepton reactions. In this context it has implications for hadron reactions. For example:

$$p + p \rightarrow p + n + \pi^+ \qquad (7.3)$$

is a possible reaction. At the quark level we have:

$$
\begin{array}{ccccccc}
p & + & p & \rightarrow & p & + & n & + & \pi^+ \\
u & & u & & u & & u & & u \\
u & & u & & u & & d & & \bar{d} \\
d & & d & & d & & d & &
\end{array}
$$

The number of u quarks in this reaction has not changed. The number of d quarks has increased by one and a $\bar{d}$ has also been produced. This would imply that the strong force has materialized a $d\bar{d}$ combination.

Notice that the d and the $\bar{d}$ *have not ended up in the same particle*. The π^+ contains one of the *original* u quarks plus the $\bar{d}$ produced by the strong force. The d that was also produced has changed places with the u in one of the protons (presumably the u that ends up in the π^+) turning that proton into a neutron.

Charge conservation is a very important rule as it allows us to say with certainty that many reactions that can be written down on paper will never be seen in experiments, such as:

$$p + p \nrightarrow p + p + \pi^+. \qquad (7.4)$$

In this reaction the total electrical charge of all the particles on the right-hand side is greater by one unit than the total charge on the left. This is not a possible hadron reaction, and has never been seen in any experiment.

If reaction (7.4) were possible, then among the quarks we would see:

$$
\begin{array}{ccccccc}
p & + & p & \nrightarrow & p & + & p & + & \pi^+ \\
u & & u & & u & & u & & u \\
u & & u & & u & & u & & \bar{d} \\
d & & d & & d & & d & &
\end{array}
$$

the total number of u quarks has increased by one and we have also attempted to produce a $\bar{d}$. The only way this reaction could take place would be by the strong force materializing a $u\bar{d}$ combination, which has a total charge of $+1$. Conservation of charge is fundamental and so the strong force is unable to do this. The reaction cannot take place.

7.2.2   Conservation of baryon number

In the last chapter we noted that all quarks have $B = 1/3$ and that all antiquarks have $B = -1/3$. This clearly implies that any materialized $q\bar{q}$ pair must contribute $B = 0$ to the total baryon number in the reaction. The consequence of this is a curious restriction on the number of particles that can be produced. To take a simple case, we again react two protons together:

$$p + p \rightarrow p + p + \pi^0 + \pi^0. \tag{7.5}$$

This reaction conserves charge, so it will take place given sufficient energy. As protons consist of three quarks they have a total baryon number of $(+1/3 + 1/3 + 1/3) = 1$, so if we examine the baryon number for the reaction as a whole:

$$
\begin{array}{ccccccccc}
 & p & + & p & \rightarrow & p & + & p & + & \pi^0 & + & \pi^0 \\
B & 1 & + & 1 & = & 1 & + & 1 & + & 0 & + & 0
\end{array}
$$

showing that baryon number works like charge conservation—the total baryon number in the reaction does not charge. On the other hand, this reaction also conserves charge:

$$p + p \nrightarrow p + \pi^+ \tag{7.6}$$

but does *not* take place as it does not conserve baryon number. Baryon number conservation effectively counts the total number of quarks involved in the reaction. The quarks can appear either as qqq ($B = 1$), $\bar{q}\bar{q}\bar{q}$ ($B = -1$) or $q\bar{q}$ ($B = 0$), no other combinations are possible. The strong force can only materialize $q\bar{q}$'s out of energy, so the number of mesons produced in the reaction is limited only by the available energy, but the net number of baryons must remain fixed (i.e. the number of baryons minus the number of antibaryons).

Consider the following list of reactions:

$$
\begin{array}{ccccccccc}
p & + & p & \nrightarrow & p & + & p & + & n & \tag{7.7} \\
p & + & n & \nrightarrow & \pi^+ & + & \pi^0 & & & \tag{7.8} \\
p & + & \pi^+ & \nrightarrow & p & + & p. & & & \tag{7.9}
\end{array}
$$

In each case charge is conserved, but the reaction is blocked by baryon number. Of course protons and neutrons are not the only types of baryon:

$$p + p \rightarrow p + \Sigma^+ + K^0 \qquad (7.10)$$

$$p + p \rightarrow p + n + K^+ + \bar{K}^0 \qquad (7.11)$$

the rule works just as well for all baryons.

You may be wondering why it is not possible to split up the qqq combinations of baryons. After all, if one can materialize q$\bar{q}$ pairs, why can't the $\bar{q}$'s produced pair up with the quarks in the baryons to produce a flood of mesons?

In order to get the $\bar{q}$'s that we would need, the strong force would have to materialize them as q$\bar{q}$ combinations. Therefore the strong force has to produce three new quarks to go with the three antiquarks that we need to break up a baryon.

The three $\bar{q}$'s could then pair off with the quarks in the baryon, decomposing it into q$\bar{q}$ mesons. But that still leaves the three quarks that were produced at the same time as the $\bar{q}$'s! These three quarks can only bind together as a qqq combination—and so we are back to square one. Hence the number of qqq combinations must remain the same.

7.2.3 Conservation of flavour

The logic of baryon number conservation applies equally to the number of different types of quark in the reaction. When a q$\bar{q}$ is materialized by the strong force, the two must be of the same flavour. Consequently the materializations cannot alter the total number of quarks of each type in the reaction.

As we commented in the previous chapter, hadrons can be labelled with internal properties that indicate the number of quarks of a certain type within them. All hadrons that contain strange quarks are referred to as strange particles and have a value of strangeness (S). The total strangeness in any strong force reaction must remain the same. For example, if we collide two non-strange particles together, then we either do not create any strange particles in the reaction, or we create a pair with opposite strangeness:

$$p + p \rightarrow p + \Sigma^0 + K^+ \quad (7.12)$$

$$S \quad 0 + 0 = 0 + -1 + 1$$

$$p + \pi^+ \rightarrow \Sigma^+ + K^+ \quad (7.13)$$

$$S \quad 0 + 0 = -1 + 1$$

By an unfortunate historical accident the K^+ was chosen as the prototype strange particle and given $S = 1$. We now know that the K^+ is a $u\bar{s}$ combination and so the s quarks has $S = -1$! Consequently the strangeness value of a hadron is equal to the number of strange quarks it contains in -1's.

This idea of conserving the individual flavours can be extended to the other heavy quarks as well: c, b and t. It has never proven useful to consider the separate conservation of u and d as they are very similar in mass.

It is important to remember that the weak force does not conserve particle types—it would allow reactions that do not conserve strangeness for example. The only situation in which this is likely to play a part is in the decay of particles (chapter 8); at other times the strong or electromagnetic forces dominate.

7.3   Using conservation laws

Conservation laws can be used to deduce information about new particles. In 1964 the following reaction was observed for the first time:

$$K^- + p \rightarrow K^0 + K^+ + X \quad (7.14)$$

Consider what we can deduce about the new particle X:

(1) X must have a negative charge—the total charge on the left-hand side is $(-1) + (+1) = 0$, as it stands the charge on the right-hand side is $(0) + (+1) = 1$ so X must have charge -1 to make up the same total.

(2) X must be a baryon—the kaons on the right-hand side are mesons so X must have baryon number $B = 1$ in order to balance the $B = 1$ from the proton.

This tells us that X must be a qqq combination.

Now, the quark contents of the particles that we are familiar with are:

K^-	+	p	$\rightarrow$	K^0	+	K^+	+	X
s		u		d		u		?
$\bar{u}$		u		$\bar{s}$		$\bar{s}$		?
		d						?

Note that the s quark in the initial K^- is unaccounted for. Quarks cannot just vanish. Hence X must contain this 'missing' s quark. One of the u quarks from the proton has gone to make up the K^- and one of the d quarks has gone to make up the K^0, but what of the other u quark from the proton?

The clue to this lies in the two $\bar{s}$ quarks that have appeared. If they have been materialized by the strong force (and where else could they come from?) then there must be two s quarks somewhere that were produced at the same time.

Where have they gone? Into X.

X is an sss combination. Two s quarks have come from the double $s\bar{s}$ materialization, and the other has come from the K^- that we started off with. So what of the missing u? Well there is one other particle unaccounted for, the $\bar{u}$ from the K^-. Evidently this $\bar{u}$ has annihilated with the u from the proton back into energy! This reaction contains two materializations and an annihilation (dematerialization!). The reaction, then is:

K^-	+	p	$\rightarrow$	K^0	+	K^+	+	X
s		u		d		u		s
$\bar{u}$		u		$\bar{s}$		$\bar{s}$		s
		d						s

This nice little exercise in logic has an extra point to it. This reaction was the first observation of the fabled Ω^- particle predicted by Murray Gell-Mann—the missing sss combination from the baryon decuplet (see chapter 6).

7.4   The physics of hadron reactions

In this section we are going to look a little deeper into the physics that lies behind the materialization and annihilation processes.

7.4.1   The field of the strong force

A force field is a volume of space inside which a force is exerted on an object. There is always another object that can be thought of as the source of the field. In the case of gravity one object, for example the earth, has a gravitational field that extends right out into space. Any object that has mass placed into this field experiences a force that attracts it towards the earth.

Despite their rather mysterious and intangible nature, gravitational fields are physically real objects. The field is a form of energy. As a mass falls towards the earth it collects energy from the field and converts this into kinetic energy. The total amount of energy that it is able to convert from the field into kinetic energy before it hits the ground is what we term the potential energy of the mass.

Electromagnetic fields act in a similar way. Any charged object is surrounded by an electric field and other charges placed within this field experience a force, the direction of which depends on the relative signs of the two charges (like charges repel, unlike charges attract). This field is slightly more tangible than the gravitational field, as it is possible to see what happens when the electric field is disturbed. If a charge is moved, then its field is disturbed. This disturbance travels through the field. If the charge is accelerating the disturbance travels as a wave—a light wave. Hence we can see what happens!

One common way of helping to visualize electrical fields is the *field line diagram*. These diagrams are a visual representation of an electrical field. Such a diagram is shown in figure 7.1.

For quarks, the closest comparable situation to the electrostatic attraction of two charges is the attraction between a quark and an antiquark. However, the properties of the strong force are sufficiently

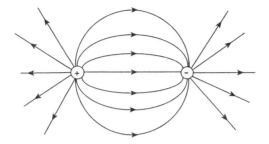

Figure 7.1 The electrical field between two attracting charges.

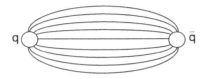

Figure 7.2 The strong field between a quark and an antiquark.

distinct to make the field line diagram very different. Figure 7.2 is an attempt at a visual representation of the strong field between a quark and an antiquark.

The field lines are more concentrated between the quarks. Hence the energy of the field is concentrated between the particles in a narrow tube-like region called a *flux tube*.

If the force between two objects is attractive, then we must supply energy in order to separate them. If we want to pick up a rock and increase its distance from the centre of the earth, then we must supply energy. The energy that we supply is transformed into field energy. We have increased the energy in the field by separating the objects. Such energy is sometimes called potential energy.

If we want to separate a quark from an antiquark we must supply energy. If we move the quark and antiquark further apart, then the flux tube connecting them gets longer. If the flux tube is getting longer there is more field, and hence more energy in the field.

In the case of the electrostatic or gravitational forces, the strength of the force decreases with distance. Specifically, if you double the distance between the charges (or masses) then the strength of the force drops to a quarter of what it was. This is known as the *inverse square law*.

If the force gets weaker with distance, then the energy needed to separate the objects decreases the further apart that they are. If two 1 coulomb charges are 1 metre apart, then to increase their separation by 1 centimetre takes 90 MJ of energy. However, if they start off 2 metres apart then to increase their separation by the same 1 centimetre only takes 23 MJ. In the case of the strong force the reverse is true. The strong force *gets stronger* the *further apart* the particles are.

If quarks are close up to each other (or q and q̄ in a meson) then there is virtually no force between them. If we try to separate them the strong force acts like an elastic band—the bigger the separation of the quarks, the greater the force pulling them back.

Separating quarks takes more and more energy the further apart they get, not less as in the electrostatic case. Consequently it is impossible to completely separate them: to do so would require an infinite amount of energy. This is the reason why the quarks can *never* be seen in isolation.

There is one problem with this. If the force between quarks increases with distance, how come there is no enormous force pulling all the quarks in my body towards one of the outer galaxies? Consider what happens when two baryons collide.

Figure 7.3(a) shows two protons moving towards each other at some speed. As a result of their collision, one of the quarks is given a tremendous thump by one of the quarks in the other proton. As a result it flies out of the proton taking a large amount of kinetic energy with it (figure 7.3(b))[1].

This quark will still be connected to the other quarks in the proton by a flux tube of strong field. As it moves away, this flux tube gets longer. In order to do this it must be gaining energy from somewhere—the obvious source being the kinetic energy of the ejected quark. As the quark moves away, its kinetic energy is being converted into strong field energy and hence into the lengthening flux tube—the quark slows

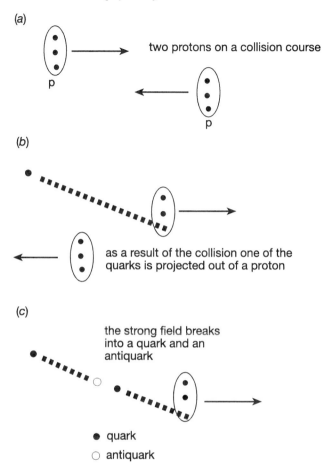

(a)

p

two protons on a collision course

p

(b)

as a result of the collision one of the quarks is projected out of a proton

(c)

the strong field breaks into a quark and an antiquark

● quark

○ antiquark

Figure 7.3 The collision of two protons leading to a reaction.

down (which is exactly what one would expect thinking of it in terms of an attractive force pulling the quark back into the proton).

Eventually, there is enough energy in the tube *to materialize a* $q\bar{q}$ *pair*. When this happens, the tube breaks into two smaller pieces (see figure 7.3(c)). The quark gets an antiquark on the other end of its piece of fractured tube, and the proton has a quark on the end of its tube. The ejected quark has turned into a meson, and the proton has gained a new quark to ensure that it remains a baryon, but not necessarily a proton. This is one physical process that lies behind hadron reactions.

How does this answer our original question about the long-range effects of the strong force?

If a quark in a distant galaxy were to interact with a quark in my body, then there would be a strong field between them. This field would be gigantic in its length, and hence would hold an enormous amount of energy. So much energy, in fact, that it would break up at myriad points along its length into quark–antiquark pairs. Hence the strong field cannot extend for very far because it becomes unstable and breaks up into mesons by materializing $q\bar{q}$ pairs. The strong force cannot extend for any great distance *precisely because it strength increases with distance.*

The flux tube mechanism can be used to explain the physical basis behind many hadron reactions. For example, if we look again at reaction (7.2):

$$p + p \rightarrow p + p + \pi^0. \tag{7.2}$$

The collision between the protons has caused one of the quarks to fly out, perhaps one of the u quarks. As it moves away the flux tube lengthens and then fractures into a $u\bar{u}$ pair, the $\bar{u}$ being on the part of the tube connected to the expelled u quark. This combination forms the π^0, while the u quark that was materialized combines with the other quarks to repair the fractured proton.

A full explanation of the physics of hadron reactions would take up more space and more technical detail than we can cover in a book of this sort. Indeed, although there are many models that are used to explain hadron reactions, a full theory has yet to be worked out. We are confident that all the physics involved is understandable in terms of the theory of the strong force, but the calculations required are beyond us at present. Simplifying assumptions have to be made. Different people argue for different assumptions, and so different models are used.

In any case, it seems clear from a study of the produced particles that slightly different mechanisms are at work depending on whether the particles are produced at the edge of the collision region (diffractive production, as it is called) or in the central 'fireball' region. The basic physics lies in the strong force materialization of $q\bar{q}$ pairs but the amount of energy and momentum involved, as well as which flavours are produced, are areas of detail that we are still working on.

7.5   Summary of chapter 7

- All reactions must conserve energy, momentum and electrical charge;
- hadron reactions conserve baryon number;
- baryon number conservation is a consequence of the limitations imposed on the materialization of new particles by the properties of the strong force;
- applying the conservation laws can help us to deduce the properties of new particles;
- fields are forms of energy that give rise to forces;
- the strong field between quarks is localized into flux tubes;
- as quarks get further apart the energy required to separate them increases;
- quarks can never be found in isolation;
- if there is enough energy in the flux tube it will fracture producing a new quark and antiquark at the ends of the pieces;
- this basic mechanism is responsible for hadron reactions.

Note

[1] This is an oversimplified model. In reality more than one quark will be jolted and there will be flux tubes connecting all the quarks together.

Chapter 8

Particle decays

In the last chapter we saw that hadron reactions are determined by the actions of the strong force. In this chapter we shall see that the weak and electromagnetic forces also have roles to play in particle decays. By studying the decays of particles we shall learn more about these forces.

Particle decay is an example of nature's tendency to reduce the amount of energy localized in an object. Nature prefers energy to be spread about as much as possible. When a particle decays the same amount of energy is shared out among a greater number of objects.

This tendency is reflected in other processes that are similar to particle decay. Radioactivity is an obvious example, but the emission of light by atoms also has features in common with decay processes.

8.1 The emission of light by atoms

When an electron is contained within an atom it is forced to adopt an exact energy value. The value is referred to as an energy level. This is a consequence of the closed nature of paths within an atom. The electron's Lagrangian restricts the energy values to ones that allow the phase to return to its initial value as you follow it round a closed path. There are many energy levels to choose from (subject to a strict set of rules), but the electron can only choose an exact level—it cannot take an energy value that falls between levels.

Sometimes an electron may be in a higher energy level than normal. When this happens the atom is said to be in an *excited state*. Atoms can be put into excited states by absorbing energy from outside, e.g. if the material is heated. The electron does not stay in this higher energy level if there is a space for it in a lower level—it 'falls' into the lower level and in so doing reduces its energy. Energy is, of course, conserved so some other object must gain energy if the electron loses it. The energy is emitted from the atom in the form of a burst of electromagnetic radiation called a photon—light is emitted.

8.2 Baryon decay

8.2.1 Electromagnetic decays

We know from chapter 6 that sometimes two distinct hadrons have the same quarks inside them and differ only in the energy levels that the quarks are sitting in.

Does this mean that we can get transitions in the energy levels of the quarks as we can with the electrons in atoms? What would we observe if such an event were to take place?

The previous discussion would suggest that we would see the emission of a photon as the drop in energy level took place. How would we go about looking for such an event?

For such an event to take place the quarks within the hadron would have to change state. If the state of the quarks is related to what type of hadron it is, as we have just recalled, then the process would result in the hadron changing into another hadron. For the moment we choose to look for such a possibility amongst the baryons.

There is a classic example that we can study:

$$\Sigma^0 \rightarrow \Lambda + \gamma. \tag{8.1}$$

This is an example of a decay equation. The object on the left-hand side of the equation was present before the decay and is replaced by the objects on the right-hand side of the equation after the decay.

We should look at the process with more care. At the quark level:

$$\Sigma^0 \quad \rightarrow \quad \Lambda + \gamma$$
$$\text{u} \qquad\qquad \text{u}$$
$$\text{d} \qquad\qquad \text{d}$$
$$\text{s} \qquad\qquad \text{s}$$

| Mass | 1.192 | 1.115 |
| (GeV/c^2) | | |

The quark contents of the two hadrons are identical. The extra mass of the Σ^0 comes from the quarks being in higher energy levels. Evidently, the decay takes place because the quarks have dropped into a lower energy state and the excess energy has been emitted in the form of a photon. This will happen spontaneously.

This decay of the Σ^0 has been closely studied, and it has been found that, on average, Σ^0 particles will decay in this way within 2.9×10^{-10} seconds of their creation in a hadron reaction. We call this an *electromagnetic* decay as a photon is emitted in the process.

If we want to find other examples then we need a hadron that has a lighter mass double into which it can decay. An obvious place to look would be amongst the Δ family of particles—the Δ^+ has the same quark content as that of a proton and the Δ^0 that of the neutron. Perhaps these particles have electromagnetic decays.

If you look them up in a data table, then you will find that about 0.6% of deltas do decay in the way that we have suggested:

$$\Delta^+ \rightarrow \text{p} + \gamma \tag{8.2}$$

$$\Delta^0 \rightarrow \text{n} + \gamma. \tag{8.3}$$

However, in most cases the deltas decay by an alternative mechanism:

$$\Delta^+ \rightarrow \text{p} + \pi^0 \tag{8.4}$$

$$\Delta^0 \rightarrow \text{n} + \pi^0. \tag{8.5}$$

Both these decays take place extremely rapidly. Typically a delta will decay some 10^{-25} seconds after it is formed[1]. Decays (8.4) and (8.5) clearly represent better options that the deltas nearly always take.

8.2.2   Strong decays

Let's look at these decays at the quark level:

$$\Delta^+ \rightarrow p + \pi^0$$

u		u	u
u		u	ū
d		d	

and

$$\Delta^0 \rightarrow n + \pi^0$$

u		u	u
d		d	ū
d		d	

As there is an obvious similarity between the two decays, we can concentrate on the first without neglecting any important physics.

Notice that the right-hand side of the quark list looks just like what we might expect in a hadron reaction *producing* the π^0. We saw something very similar when we looked at the reaction $p + p \rightarrow p + p + \pi^0$ on page 122 of chapter 7. In that reaction we discovered that the strong field had been disturbed by quarks colliding. The field then used the energy it gained to materialize a quark/antiquark pair, which went on to form the pion.

If something similar is happening in (8.4) the strong field inside the Δ^+ must have enough energy to materialize the uū pair. We can check that by looking at the masses of the Δ^+ and the proton:

$$\text{p mass} = 0.938 \text{ GeV}/c^2$$
$$\pi^0 \text{ mass} = 0.135 \text{ GeV}/c^2$$
$$\overline{\text{total} = 1.073 \text{ GeV}/c^2}$$
$$\Delta^+ \text{ mass} = 1.23 \text{ GeV}/c^2.$$

On this basis there looks to be more than enough energy! The total mass of the proton and a π^0 is much less than that of a Δ^+. This is an example of a *strong decay*.

Inside a hadron, the quarks are bound to each other by the strong force. Hence the quarks within the hadron are wading about in a strong field.

The strong force is very much stronger than the electromagnetic forces between the quarks and so it is more significant in determining their energy levels. When the quarks lose energy a disturbance is set up in the strong field—the energy is localized in a certain part of the field (I visualize it to be like a lump in some custard, the custard representing the strong field). This disturbance is emitted in a similar way to the photon in the electromagnetic case. A moving packet of disturbance in the strong field is called a *gluon* (pronounced 'glue-on'). Gluons are to the strong force what photons are to the electromagnetic force.

The analogy between photons/electromagnetic field and gluons/strong field is a good one and has proven very useful in working out the theory of the strong force.

We conclude that the quarks in the Δ^+ are able to switch energy levels by emitting a gluon. However, the strong force has a trick up its sleeve. Out of the energy of the gluon materializes a quark and an antiquark, in this case either a $u\bar{u}$ or a $d\bar{d}$ that form the π^0.

Physicists like to draw sketch diagrams such as figure 8.1 to illustrate such processes.

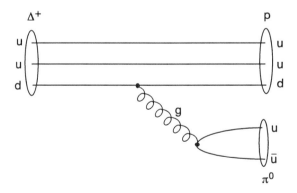

Figure 8.1 The Δ^+ decay drawing.

In this drawing the complete lines represent quarks that are 'moving' through the diagram, in the sense that the left-hand side of the diagram represents the initial state (the Δ^+ in this case) and the right-hand side represents the final state. The loops at each end show that the quarks are bound into hadrons, and the specific hadron is labelled above the loop. The curly line is the gluon. As you can see, the diagram shows a quark in the Δ^+ emitting a gluon which then materializes into a $u\bar{u}$ combination that forms a pion.

Be careful not to read too much into this diagram. The lines are *not* supposed to represent definite paths of the particles. It is just a sketch to help visualize what is happening.

There is another way in which the Δ^+ can decay via the strong force:

$$\Delta^+ \rightarrow n + \pi^+. \tag{8.6}$$

This is a slightly more complicated case. We can explain it by presuming that the gluon materializes into a $d\bar{d}$ combination, but that they do not go on to bind to each other; instead the d joins with the u and one other d from the initial Δ^+ to form the neutron, and the $\bar{d}$ joins with the leftover u from the Δ^+ to form the π^+. One of our diagrams (figure 8.2) will help in visualizing this.

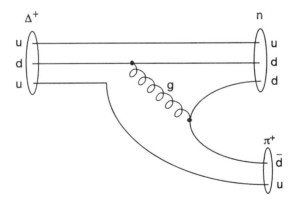

Figure 8.2 An alternative Δ^+ decay.

The rest of the delta family of hadrons can also decay:

$$\Delta^{++} \rightarrow p + \pi^+ \tag{8.7}$$

$$\Delta^- \rightarrow n + \pi^- \tag{8.8}$$

with similar lifetimes to the decays that we have been discussing.

Returning to the Σ^0 decay, it is interesting to see why the decay cannot proceed via the strong force. After all, there does appear to be a process that conserves both charge and strangeness open to it:

$$\Sigma^0 \rightarrow \Lambda + \pi^0. \tag{8.9}$$

In order to see the answer we must consider the masses of the particles concerned.

$$\Lambda \text{ mass} = 1.115 \text{ GeV}/c^2$$
$$\pi \text{ mass} = 0.135 \text{ GeV}/c^2$$
$$\text{total} = 1.250 \text{ GeV}/c^2$$
$$\Sigma^0 \text{ mass} = 1.193 \text{ GeV}/c^2.$$

The total mass of a Λ and a π^0 is *greater* than the mass of a Σ^0. The strong decay is blocked as there is not enough excess energy in the Σ^0 to create the pion out of the strong field, hence the decay proceeds via the electromagnetic force.

We can divide the particles that we discussed in chapter 6 into groups depending on the manner of their decay. We have two sets of baryons to deal with (see tables 8.1 and 8.2), the decuplet and the octet. The

Table 8.1 The baryon decuplet.

Particle family					Average mass (GeV/c^2)
Δ^-		Δ^0		Δ^+		Δ^{++}	1.23
	Σ^{*+}		Σ^{*0}		Σ^{*-}		1.39
		Ξ^{*0}		Ξ^{*-}			1.53
			Ω^-				1.67

Table 8.2 The baryon octet.

Particle family					Average mass (GeV/c^2)
	p		n		0.939
Σ^+		Σ^0/Λ		Σ^-	1.19/1.12
	Ξ^0		Ξ^-		1.32

octet is the set of lower mass baryons some of which are reflected in higher mass versions that lie in the decuplet.

We have already dealt with the delta decays, but now we see that the high mass particles in the decuplet can decay into low mass versions in the octet. For example:

$$\Sigma^{*+} \rightarrow \Sigma^+ + \pi^0 \tag{8.10}$$

$$\Sigma^{*+} \rightarrow \Lambda + \pi^+ \tag{8.11}$$

$$\Sigma^{*0} \rightarrow \Sigma^0 + \pi^0 \tag{8.12}$$

are all possible decays. Note that the Σ^{*0} has a strong decay open to it which it will take in preference to the electromagnetic one (in this case there is a big enough mass difference). I am sure that you can make up other possible decays for the particles in the decuplet. Just to emphasize the point here are a few more:

$$\Xi^{*-} \rightarrow \Xi^- + \pi^0 \tag{8.13}$$

$$\Xi^{*0} \rightarrow \Xi^0 + \pi^0 \tag{8.14}$$

$$\Xi^{*0} \rightarrow \Xi^- + \pi^+. \tag{8.15}$$

However, when we come to the Ω^- we hit a problem. There is no member of the octet with the same quark content (sss) and so there is no strong or electromagnetic decay open to it. The Δ^{++} (uuu) and the Δ^- (ddd) can decay by the strong force using a mechanism (figure 8.3) similar to one we have drawn before and it is possible to draw a similar diagram for the Ω^- (figure 8.4).

However, this decay is blocked as there is not enough mass:

$$\Omega^- \rightarrow \Xi^- + K^0$$

$$\Sigma^- \text{ mass} = 1.197 \text{ GeV}/c^2$$

$$\underline{K^0 \text{ mass} = 0.498 \text{ GeV}/c^2}$$

$$\text{total} = 1.695 \text{ GeV}/c^2$$

$$\Omega^- \text{ mass} = 1.672 \text{ GeV}/c^2.$$

Gell-Mann was aware of this when he predicted the existence of the Ω^- and that it would have a lifetime longer than those of the other members of the decuplet. As we have seen, the strong decays typically take place in about 10^{-25} seconds, whereas the Ω^- typically takes 1.3×10^{-10} seconds to decay.

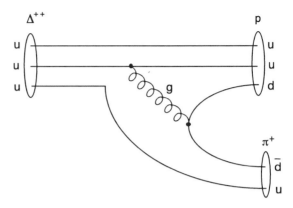

Figure 8.3 The Δ^{++} strong decay.

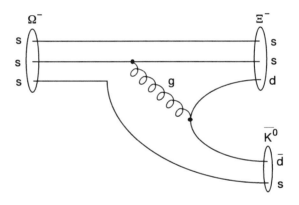

Figure 8.4 A possible Ω^- decay.

8.2.3   Weak decays

Before reading this section it would be wise to review the section of chapter 7 called 'the field of the strong force'. We are going to use the ideas of a field quite a lot in this section, so you ought to be sure that you are reasonably happy with what we have said so far on the subject before carrying on.

At the end of the last section we left open the question of how the Ω^- particle was able to decay as the strong and electromagnetic channels were blocked to it. The same is true for the majority of particles in the baryon octet. The Σ^0 can decay electromagnetically into the Λ but there are no light mass versions of the Σ^+ for example (the octet is after all the lowest mass versions of these quark contents), so how can these particles decay?

One of the characteristics of the weak force is that it allows quarks to change flavour, and that is what is needed here. If for example the Σ^+ were to decay, the only particles lighter than it would be the proton and neutron, neither of which contains a strange quark. The strange quark has to be removed in order for the Σ^+ to decay.

The Σ^+ has several possible decays open to it, but the most common ones are:

$$\Sigma^+ \rightarrow p + \pi^0 \tag{8.16}$$
$$\Sigma^+ \rightarrow n + \pi^+. \tag{8.17}$$

Let us look at these at the quark level to see if we can deduce what is going on:

$$
\begin{array}{ccc}
\Sigma^+ & \rightarrow p & + \pi^0 \\
u & u & u \\
u & u & \bar{u} \\
s & d &
\end{array}
$$

It is obvious from this that two things have happened: the s quark has disappeared and we have picked up u, $\bar{u}$ and d quarks instead. Evidently the s has been turned into one of these other quarks and two more have appeared—probably out of the field as in the case for strong decays.

If this decay is to follow the pattern that we have already established for the strong and electromagnetic decays then there must be some sort of transition from a high energy state into a low energy state coupled by the emission of some disturbance to the field. Obviously in this case the transition also entails turning an s quark into some other flavour—either u, d or $\bar{u}$ according to the quarks that turn up on the right-hand side of the decay equation.

Years of work and study of the processes of weak decays and interactions have established that the weak force must follow a set pattern with regard to the quarks. The evidence for this pattern is compelling but complicated, so I am going to have to ask that you accept it as one of the rules. The weak force acts on the quarks in the manner outlined in table 8.3.

Table 8.3 The action of the weak force on the quarks and leptons.

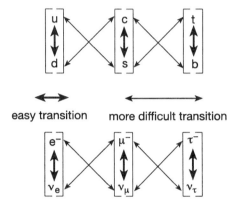

Table 8.3 is telling us is that it is possible for the weak force to change the s quark into a u quark. This is obviously a beneficial transition to take place within the Σ^+ as the u quark is less massive, and therefore the energy of the hadron will be reduced. This is not quite the same as a quark dropping energy levels, in a sense it is more fundamental than that. In this case the quark is changing into a different particle and in the process giving up some energy to the weak field.

A quark has an electrical charge and consequently is always surrounded by an electromagnetic field. Similarly, a quark is always surrounded by strong and weak fields.

Within the Σ^+ the s changes into a u producing a disturbance in the weak field. This disturbance is similar to a gluon or a photon in some ways, but there is a very important difference: it is electrically charged[2].

Just putting to one side for the moment what that actually means, let us see *why* it must be so.

If an s quark ($-1/3$) can change into a u quark ($+2/3$) by giving energy over to the weak field, then it must give over charge as well. If we say that the field gains a charge Q, then conservation of charge suggests:

$$\text{initial charge (s quark)} = -1/3$$
$$\text{final charge (u quark)} = +2/3 + Q$$

hence

$$-1/3 = +2/3 + Q$$
$$\therefore \quad Q = -1.$$

When a quark changes state in a weak field it gives up both energy and charge to the field, and the object that is produced is called a 'W'.

Actually, there are two types of W, the W^+ and the W^-. It is clear that when the s changes to a u quark it must emit a W^- in order to balance the charge change. Similarly, if a c were to change into a d, then it would have to emit a W^+.

Particle physicists like to draw sketches (as in figure 8.5) to help them remember this.

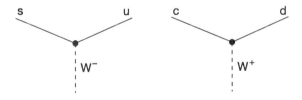

Figure 8.5 Some of the weak transitions of quarks.

If you find it difficult to remember then you can always work it out by balancing the charges (that's what I do—I can never remember it!).

So far we have said that the s turns into a u by emitting a W^-, but what then happened to this? In the case of the strong decay the gluon materialized a q$\bar{q}$ pair. The same thing happens to the W^-, but true to the nature of the weak force, the quark and antiquark do not have the same flavour. In this case, the W⁻ materializes into a dū pair. The gluon is not electrically charged, so it is only able to materialize quarks/antiquark pairs of the same flavour.

In order to summarize all this, we will draw one of our sketches, see figure 8.6.

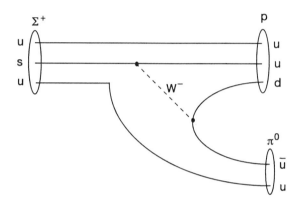

Figure 8.6 The Σ^+ decay.

8.2.4 Neutron decay

The neutron is an unstable particle when it is in isolation. Inside the nucleus, the effect of the other protons and neutrons is to stabilize the individuals so that they can last indefinitely. If this stabilizing effect is not quite complete, then the result is a form of β radioactivity.

The decay process is:

$$n \rightarrow p + e^- + \bar{\nu}_e \tag{8.18}$$

or at the quark level

$$d \rightarrow u + e^- + \bar{\nu}_e. \tag{8.19}$$

The existence of this process was well known in the 1930s, but it presented a profound difficulty to the physicists of the time. The existence of the $\bar{\nu}_e$ was unsuspected. Only the emitted e⁻ could be

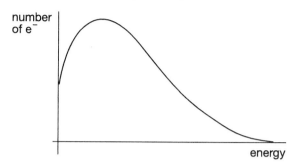

Figure 8.7 The kinetic energy distribution of e⁻ produced in β decay.

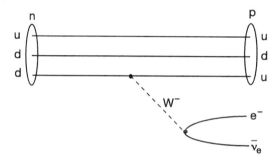

Figure 8.8 Neutron decay.

detected. Measurements of the e⁻ energy showed that it was produced with a wide range of energy values up to some maximum (figure 8.7).

This seemed to violate the law of conservation of energy which would suggest that as there were only two objects involved after the decay (the proton and the electron) the split of the available energy between the two of them should be fixed—hence the e⁻ should always have the same energy. Wolfgang Pauli suggested that the puzzle could be solved if another neutral particle was present after the decay making the split of energy a three-way process and hence not fixed. It was not until 1953 that the neutrino was discovered (see page 77).

Neutron decay (figure 8.8) is a typical weak decay which is even slower than normal as the proton and neutron have only a very small mass difference between them.

8.3 Meson decays

As the W's can materialize into q$\bar{q}$ pairs of different flavour, it should not surprise us to discover that the weak field can also be involved in the annihilation of quarks of different flavours.

Direct evidence of this process can be seen in the decay of the lightest mesons. The simplest family of mesons is the pion family π^+, π^- and π^0.

The π^+ and the π^- are antiparticles of each other containing u$\bar{d}$ and d$\bar{u}$ quark combinations respectively. Conservation of electrical charge and flavour forbids the pions to decay via the strong or electromagnetic forces, so a weak decay is the only avenue open to them.

We already know from the Σ^+ decay that the W^- is capable of materializing into a d$\bar{u}$ quark combination. In the case of the π^- the same quark combination is capable of mutual annihilation by turning into a W^-.

As the d and $\bar{u}$ quarks are amongst the lightest, there is no advantage in the π^- decaying into a W^- which then materializes quarks again. The only other possibility is for the W^- to turn into a lepton pair:

$$\pi^- \rightarrow \mu^- + \bar{\nu}_\mu. \tag{8.20}$$

We can sketch this process as in figure 8.9.

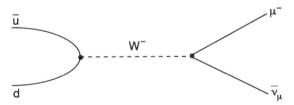

Figure 8.9 The π^- decay.

This is just as much an annihilation reaction as the e^+e^- reactions and the q$\bar{q}$ $\rightarrow$ g reactions that we have studied. In this case it is the weak field which is excited by the annihilation rather than the electromagnetic or the strong.

Notice that the decay does conserve both baryon number ($B = 0$ initially) and lepton number.

The weak field is therefore capable of a complex set of reactions. It can either cause flavour transformations, annihilations or materializations. It is possible to summarize all the possibilities in a set of sketches (see figure 8.10).

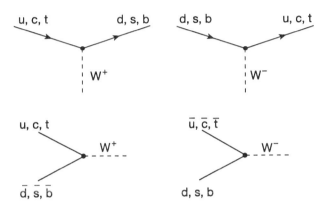

Figure 8.10 The W couplings to the quarks and antiquarks.

These sketches show the ways in which quarks can 'couple' to the weak field. Notice that the lower two sketches in the set indicate that the $q\bar{q}$ pairs listed can annihilate into a W, or in reverse the W can materialize into the pair. Remembering all these couplings is not quite as daunting as it seems. Once you have mastered the quark generations simple conservation of charge enables one to work out the possibilities.

The third member of the pion family, the π^0 also decays, but as it has the quark content $u\bar{u}/d\bar{d}$ it can decay electromagnetically—matter and antimatter annihilating into photons (see figure 8.11):

$$\pi^0 \rightarrow \gamma + \gamma. \qquad (8.21)$$

Notice that conservation of energy and momentum forbids the π^0 from decaying into a single photon. Inside the pion a continual conversion takes place involving the exchange of two gluons (see figure 8.12), which is why the π^0 is both a $u\bar{u}$ and a $d\bar{d}$ combination.

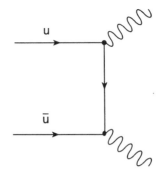

Figure 8.11 The π^0 decay.

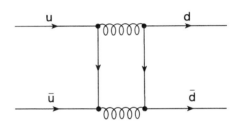

Figure 8.12 Inside a π^0.

8.4 Strangeness

In the early days of hadron reactions, it became clear that there was a class of particle that could be produced quite easily in the reactions, were quite massive and yet took comparatively long times to decay. The fact that they were produced quite easily suggested that they were being created by the action of the strong force, making them hadrons. If this was true, then it was puzzling that they did not seem to decay via the strong force as well. This was clearly not the case as they had lifetimes typically of the order of 10^{-10} s.

The other conspicuous feature was that these particles were only ever created in pairs e.g.:

$$p + p \rightarrow p + \Sigma^+ + K^0 \qquad (8.22)$$

but never singly,

$$p + p \nrightarrow p + \Sigma^+ + \pi^0. \tag{8.23}$$

This feature led Gell-Mann, Nakano and Nishijima in 1953 to suggest that the whole puzzle could be solved by assuming the existence of a new internal property of hadrons. They named this property *strangeness* (after the odd behaviour of the particles) and suggested that it must be conserved in strong interactions. This explained why the particles were produced in pairs. In reaction (8.22) the protons both have strangeness $S = 0$, so the initial total is zero. After the reaction, the Σ^+ has $S = -1$ and the K^0 has $S = +1$ making the total zero again.

When the particles come to decay, they can decay into another strange particle if there is a less massive one in existence e.g.:

$$\Sigma^{*+} \rightarrow \Sigma^+ + \pi^0 \tag{8.10}$$

which would be a fast strong decay. However, if there was no lighter strange particle available then the strong force could not be used as the decay would have to involve a loss of strangeness;

$$
\begin{array}{ccccc}
\Sigma^+ & \nrightarrow & p & + & \pi^0 \\
S \quad -1 & \neq & 0 & + & 0.
\end{array}
$$

The suggestion was that the weak force did not conserve strangeness and hence would be the force involved in the decay. Hence the lifetime would be much longer. This suggestion of a new internal property put Murray Gell-Mann and others on the road to the classification of hadrons that we discussed in chapter 6 (the eightfold way) and eventually to the discovery of quarks.

8.5 Lepton decays

Everything that we have said about the decay of hadrons can be applied to the decay of leptons. We already know from chapter 4 that some leptons can decay. For example:

$$\mu^- \rightarrow e^- + \nu_\mu + \bar{\nu}_e. \tag{5.2}$$

This decay proceeds via the emission of a W particle by the μ^- which changes it into a ν_μ (see figure 8.13).

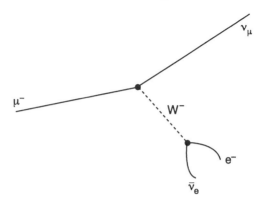

Figure 8.13 The decay of the μ^-.

Tau decay takes place in a similar manner, but electrons are prevented from decaying by conservation of lepton number.

8.6 Summary of chapter 8

- Particles decay in order to rid themselves of excess energy;

- a simple electromagnetic decay is caused by quarks dropping into lower energy levels and emitting photons;

- an electromagnetic decay has a lifetime typically of the order of 10^{-20} seconds;

- strong decays take place when quarks drop into lower energy levels and emit gluons;

- the emitted gluon then materializes into a $q\bar{q}$ pair;

- strong decays have lifetimes typically of the order of 10^{-25} seconds;

- a weak decay takes place when a quark/lepton changes into another quark/lepton by emitting a W^+ or a W^-;

- the W's then materialize into a $q\bar{q}$ pair or a lepton and an antilepton (conserving lepton number);

- weak decays have lifetimes typically of the order of 10^{-10} seconds.

Note

[1] This is such a short time that most people refuse to dignify them with the name *particle*, preferring to call them *resonances*. I will continue to call them particles so as to save introducing a new term.

[2] There is also a neutral weak force particle called the Z^0, see section 11.2.2.

Chapter 9

The evidence for quarks

The evidence for the existence of quarks has built up over the last thirty years. Physicists have taken some time to be convinced by the various threads of evidence. In this chapter we shall look at the various arguments that have been put forward to justify the belief in quarks.

9.1 The theoretical idea

In 1964 Gell-Mann and Zweig independently came up with the idea that the 'eightfold way' pattern of arranging hadrons (see chapter 6) could be explained by taking three basic sets of properties and combining them in various patterns[1]. As a mathematical structure the idea was very pleasing but there was no direct experimental evidence for it and it did raise some disturbing theoretical problems, so it was not widely accepted. However, the search for elegance is a powerful motivator in physics and this was certainly an elegant idea, so it did attract some attention.

9.2 Deep inelastic scattering

If physicists required some more direct evidence for the existence of quarks then it was not long in coming. The first real evidence came as a result of the analysis of data taken by an experiment at the Stanford Linear Accelerator Centre (SLAC) in the late 1960s. The leaders of this experiment, Friedman, Taylor and Kendal, were awarded the 1990 Nobel Prize in Physics. The Stanford experiment was, in essence, a generalization of a famous experiment performed by Geiger and Marsden in 1909. As the first person to produce a theoretical explanation for their results was Ernest Rutherford, the physics involved has become known as 'Rutherford scattering'.

9.2.1 Rutherford scattering

When this experiment was first performed physicists were using a model of the structure of the atom that had been proposed by J J Thomson in 1903. Thomson suggested that the negatively charged electrons, that were known to be present inside atoms, were embedded into a positively charged material that made up the bulk of the atom (protons had not been discovered). The atom was a solid ball into which the electrons were stuck, rather like the plums in a pudding. Thomson's idea became known as the 'plum pudding' model of the atom.

As the nature of the positive material was unknown, Rutherford was keen to find out whether the positive 'stuff' was uniformly spread throughout the volume of the atom, or distributed in some other fashion. Earlier experiments had shown him that α particles could be deflected through large angles when passed through thin sheets of mica. However, Rutherford was unable to produce similarly large deflections by applying very strong electrical or magnetic fields directly to α particles. Clearly the forces at work within the mica sheets were far greater than those that Rutherford could produce in the lab.

In 1909 Rutherford suggested to Ernest Marsden, a student of Rutherford's collaborator Hans Geiger[2], that he use α particles from a naturally radioactive isotope to try to determine how big a deflection could be produced. Rutherford was hoping to find the distribution of positive charge within the atom by studying the ways in which the positively charged α particles were deflected by electrostatic repulsion.

Marsden elected to use a thin foil of gold as the target for the α particles. Gold is a relatively heavy atom and the metal can easily be formed into thin sheets. The thinner the target, the smaller the number of atoms that the α particles would collide with as they passed through, making the collisions easier to study. (α particles do not penetrate very far through matter, so the foil had to be thin for another reason—the α particles would never make it through to the other side if it was any thicker!)

The result of Marsden's experiment came as a total surprise to Rutherford and Geiger. Marsden discovered that most of the α particles passed through the foil and were hardly deflected. However, approximately 1 in 20000 were deflected to such an extent that they reversed course and bounced back in the direction that they had come from. Rutherford described this as being like firing a 15 inch shell at a piece of tissue paper and watching it bounce off.

Although only a small proportion of the α particles were bounced back, calculations convinced Rutherford that positive material distributed throughout the atom could not be concentrated enough to exert such a large force on an α particle. This puzzle worried Rutherford for at least a year until in 1910, with the aid of a simple calculation, he saw the significance of Marsden's results.

He reasoned that if an α was to be turned right back, then it must come to rest at some distance from the centre of the atom and then be pushed back. If it came to rest then all its kinetic energy must have turned into electrostatic potential energy at that distance. Knowing the amount of kinetic energy the α had to start with it was a simple matter to calculate the distance. The answer astounded him. The α particles that were completely reversed had to come within 10^{-15} m of the *total* positive charge of the atom. As the diameter of a gold atom was known to be of the order of 10^{-10} m, this implied that *all* the positive charge was concentrated in a region at the centre of the atom some hundred thousand times smaller than the atom itself. Thomson's model of the atom could not have been more wrong[3].

Rutherford used this result as the justification for his nuclear model of the atom. He suggested that all the positive charge is concentrated in a very small region, known as the nucleus, in the centre of the atom. The electrons were in orbit round this region, rather than embedded within it, in a similar way to that in which the planets orbit the sun. We now appreciate that this crude model is inadequate in many important ways[4], but the essential feature of the nucleus is retained.

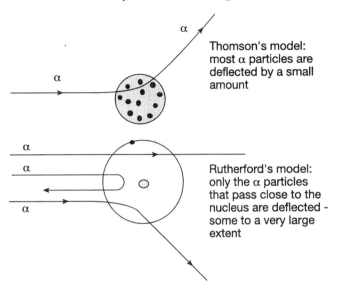

Figure 9.1 Contrasting models of the atom.

Figure 9.1 contrasts Thomson's and Rutherford's atomic models. It was not long before Rutherford's model was refined by the discovery of the proton and neutron inside the nucleus. This, in essence, is the model that we still use today.

9.2.2 The SLAC experiment

The SLAC experiment worked on the same basic theme as Rutherford scattering: using a particle to probe the structure of an object. In this case, however, the probe particles were electrons accelerated to very high energies and the object was the proton.

Electrons are significantly easier to use than α particles. They are easily produced in large numbers; they have low mass, and so can be accelerated comparatively easily, and they do not feel the strong interaction.

The last point is significant. The Stanford physicists intended to fire high energy electrons at protons and to measure the extent to which they were deflected to see how the electrical charge is distributed throughout the volume of the proton. If they tried to use a hadron as the probe, then it would react with the proton via the strong interaction. Such reactions are much harder to analyse than those caused by the electromagnetic force. In any case, the strong force does not depend on the electrical charge of the objects concerned, so the resulting reactions would not provide any direct information about the charge distribution within the proton.

What they found took them by surprise.

At low energy the electrons deflected slightly and the protons recoiled. As they increased the energy of the electrons a threshold energy was reached beyond which the electrons were deflected through much greater angles. At the same time the protons started to fragment into a shower of particles rather than being deflected themselves. They had started to see evidence for charged quarks inside the proton.

9.2.3   Deep inelastic scattering

At the time, the results of the SLAC experiments were a great puzzle— the quark theory was not well known. The full theory that explained the results was developed by Feynman and Bjorken in 1968. Feynman named the small charged objects inside the proton 'partons'. However, it soon became clear that Feynman's partons were the same as Gell-Mann's quarks.

(Some people maintain a distinction by using the name parton to refer to *both* the quarks and the gluons inside protons and neutrons.)

The effect of a high speed charged object passing close to the body of another charged particle, such as a proton, is to produce a severe disruption in the electrostatic field that connects the particles. The exact nature of this disruption depends on the motion of each of the charges involved. At a crude level of approximation we can think of this disruption as being caused by the passage of a photon from the electron to the proton. We can draw one of our sketches to represent this process (see figure 9.2).

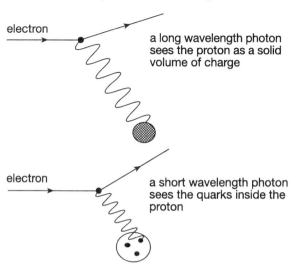

Figure 9.2 The interaction between a photon and a proton.

It is important to have some physical picture of what is happening in this process. Many authors have suggested various ways of looking at this, none of which seem entirely satisfactory to me. I wish to base my account on the very physical picture used by Feynman.

The emission of a photon by an electron is not independent of its subsequent absorption. Diagrams like figure 9.2 encourage us to think of the photon as a particle moving through empty space (as symbolized by the wavy line). In truth it is more like the motion of a ruck in a carpet that is being pulled tight. The carpet is important in determining the motion of the ruck. Although we do not draw the electromagnetic field on our diagram, it is there and it connects the charged objects (it is the 'carpet'). Whether it is a high or low energy photon that is passed depends on the motion of *both* charges and the consequent disturbance of the *total* field.

A low energy electron is moving comparatively slowly and so takes some time to pass by the proton. In this time the quarks within the proton, which are moving at a very high speed, cover considerable distances although they remain localized within the proton. Hence the electron 'sees' an average electromagnetic field that is smeared out over the volume of the proton due to the motion of the quarks. (Do not forget that the charge of the proton is simply the sum of the charges of the quarks inside.) This 'smeared' field is of relatively low intensity and so the interaction only triggers a low energy photon from the electron. If the electron emits a low energy photon then it is hardly deflected from its path and the photon, which has a long wavelength, is absorbed by the whole proton.

On the other hand, a high energy electron passes the proton very quickly. In the time it takes to pass the proton the quarks hardly move[5] and so the electromagnetic field has several localized intense regions (corresponding to the positions of the quarks). An electron passing close to one of those regions will cause a severe localized disturbance in the field—a high energy photon is emitted and then absorbed by the quark.

Imagine a set of helicopter blades. An object passing by the blades slowly would see them as an apparently solid disc. A quick object would see individual blades as they do not have the chance to move very far while the object passes.

If the electron emits a high energy photon (short wavelength) it is deflected through a large angle. The photon is absorbed by an individual quark within the proton. As a result of this the quark is kicked out of the proton as well. The quark will move away from the proton, stretching the strong field until it fragments into a shower of hadrons. The result is a deflected electron and a shower of hadronic particles.

When the experimental results show that the electrons are being deflected by large angles and the proton is fragmenting into a shower of hadrons, then we can deduce that the electron is interacting directly with the quarks within the proton.

The SLAC experiment ran for many years and was refined by using other particles, such as neutrinos and muons as probes. The SLAC people were able to establish that there are three quarks inside the proton and that their charges correspond to those suggested by Gell-Mann and Zweig. Deep inelastic scattering forms the cornerstone of our physical evidence for the existence of quarks.

9.3 Jets

The third major piece of evidence to support our belief in the existence of quarks comes from the study of e^+e^- annihilation reactions. We first mentioned annihilation reactions in chapter 5 when we saw that they could be used to produce new flavours of leptons. The basic reaction takes the form

$$e^+ + e^- \rightarrow \mu^+ + \mu^- \tag{9.1}$$

a reaction that becomes possible once there is enough energy in the incoming electron–positron pair to materialize the muon and the antimuon. Since then we have looked at similar processes, hence we recognize that the electromagnetic force is taking energy from the annihilation and is then materializing new particles from that energy. We can even draw a sketch diagram to illustrate the process (see figure 9.3) in which the photon represents the energy flowing into and out of the electromagnetic field.

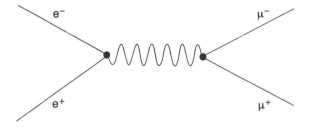

Figure 9.3 A typical annihilation reaction.

Evidently the electromagnetic force has a free hand when it comes to materializing the new particles. Baryon number, lepton number and electrical charge are all zero initially, so as long as opposites are created, then anything can come out the other end. Specifically the reaction:

$$e^+ + e^- \rightarrow q + \bar{q} \tag{9.2}$$

is perfectly possible, provided the q and the $\bar{q}$ are the same flavour (i.e. u$\bar{u}$, d$\bar{d}$, etc).

In the case of leptons being materialized the reaction is perfectly simple: the lepton and antilepton are produced and move away from each other along the same line (to conserve momentum). It is not so simple for quarks because of the strong force. The strong force will not allow quarks to appear on their own. It is very unlikely that the two quarks will bond to form a meson, as they are produced moving away from each other. They will be linked by a strong force flux tube (see chapter 7) that stretches as they move apart. The quarks will lose kinetic energy as the flux tube gains it. Eventually the flux tube will break up into mesons. The result will be two showers of particles emerging from the reaction point—largely in the direction of the original quarks. There will be some divergence; the particles will not appear along two ruler-like lines, but they do retain much of the initial quark's motion. Appropriate equipment can record their paths producing quite striking computer displays such as that shown in figure 9.4.

The streams of particles have been christened *jets* and they are striking visual evidence for the existence of quarks.

There is even more to be learned from the study of jet events.

In chapter 5 we discussed the various threshold energies that are passed as the e^+e^- energy is increased. Each threshold corresponds to there being enough energy to produce a new, more massive, flavour of particle. A threshold is reached at each new quark or lepton mass. This provides a nice way of counting the numbers of quarks that exist.

There is no problem in counting leptons. When they are produced they are not accompanied by jets of particles—they can be seen and studied in isolation. Quarks are a different matter. There is no direct way of telling what flavour of quark is produced by looking at the jets of particles. Their identities are completely swamped by the other hadrons. There is, however, a beautiful way of flavour counting—one simply looks at the relative number of times hadrons are produced as opposed to leptons.

Specifically, physicists study the ratio R:

$$R = \frac{\text{prob}(e^+e^- \rightarrow \text{hadrons})}{\text{prob}(e^+e^- \rightarrow \mu^+\mu^-)}.$$

We can see that the ratio will increase every time a new flavour threshold is passed. All the reactions that were possible below the

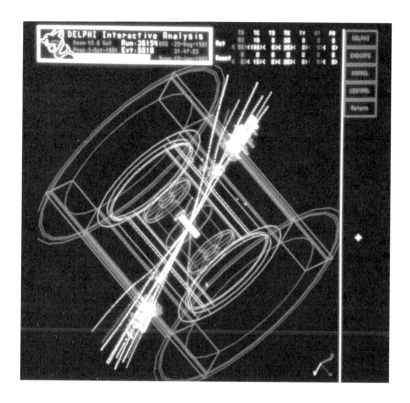

Figure 9.4 A two-jet event seen in the DELPHI detector (see chapter 10). (Courtesy of the CERN press office.)

threshold still continue, but above the threshold there is another reaction possible that will increase the number of events that produce hadrons. Notice that we do not worry about *what* hadrons—that the annihilation produces hadrons *at all* is an indication that a q$\bar{\text{q}}$ pair has been formed.

R depends only on the quark charges. Specifically:

$$R = \Sigma(Q_i^2)$$

in which the summation only counts the charges of those quarks whose masses are low enough to be produced at the specific energy.

R is measured at various energies. If the energy is such that only the u and d quarks can be produced, then R should take the value

$$R = (+2/3)^2 + (-1/3)^2 = 5/9.$$

This should then stay constant until the energy is great enough for the s quark to be produced, at which point

$$R = (+2/3)^2 + (-1/3)^2 + (-1/3)^2 = 2/3.$$

Then again, once the c threshold is passed

$$R = (+2/3)^2 + (-1/3)^2 + (-1/3)^2 + (+2/3)^2 = 10/9$$

and so on. Above 10 GeV the u, d, s, c and b quarks are all produced giving an R value of $11/9 = 1.22$. The actual value, as measured by experiments at the PETRA accelerator in Germany, is 3.9 ± 0.3 which is clearly in disagreement with the theory. The solution to this problem lies in another property of quarks which turns out to be the key to understanding the strong force.

9.3.1   Colored quarks

Quarks feel the strong force, whereas the leptons do not. Evidently there is a property of quarks that the leptons do not possess. Baryon number is the obvious candidate, but this does not explain why quarks must bond together as qqq, $\bar{q}\bar{q}\bar{q}$ or $q\bar{q}$ combinations. The correct suggestion was made by Greenberg, Han and Nambu, who realized that there is a strong force 'charge' comparable to the electrical charge that a particle must have to experience the electromagnetic force. Electrical charge comes in two varieties: positive and negative. Their property had to come in three types if it was to explain quark bonding. They named the property *color*[6].

Greenberg, Han and Nambu theorized that quarks can come in three colors: red, blue and green and that any stable hadron must be a colorless combination of quarks. Hence a baryon must contain a red, blue and green quark. A proton, for example, could be u(red)u(blue)d(green), or u(r)u(g)d(b), or u(g)u(b)d(r), etc. Any combination will do as long as there is only one of each. Antiquarks have anticolor: antired, antiblue and antigreen. A meson is therefore a colorless combination of color and anticolor. For example a π^+ would be u(r)d̄(r̄), or a green–antigreen or a blue–antiblue. From this unlikely beginning the whole, successful, theory of the strong force was developed.

Obviously color is not meant in the visual sense. The strong force charge was named color as the pattern of combining three quarks of different charge together is reminiscent of combining primary colours together to make white. The only connection lies in the name. Think of color as being the strong force version of electrical charge.

The relevance of this to our discussion of R is that it multiplies each R value by 3. Whenever the reaction

$$e^+ + e^- \rightarrow q + \bar{q} \tag{9.2}$$

takes place, the resulting quarks could be either red/antired or blue/antiblue or green/antigreen. For each quark flavour there are three distinct quarks that can be produced. The reaction is three times more likely.

If we multiply the 10 GeV prediction by three, then we get $11/3 = 3.67$ which is well within the experimental uncertainty of the measured value. Not only does color form the basis of a theory of the strong force, but it is obviously required to make the theoretical R values match the measured values.

9.4   The November revolution

When the quark hypothesis was first put forward only three quarks, the u, d and s, were required to explain all the hadrons that had been discovered. Hence the list of elementary particles known to physicists comprised u, d and s quarks along with e^-, ν_e, μ^- and ν_μ; three quarks and four leptons. This apparent asymmetry between the number of leptons and quarks led Glashow and Bjorken to suggest, in 1964, that there might be a fourth quark to even up the numbers. They named this hypothetical object the charm quark. At the time there was no experimental evidence to justify such an extension. Later Glashow, Iliopoulos and Maini used the idea to provide an explanation for the non-occurrence of the decay $K^0 \rightarrow \mu^+ + \mu^-$. The explanation is technical and beyond the scope of this book, but it did something to legitimize the suggestion of a fourth quark.

The real revolution came in November 1974 when two teams of experimentalists announced the independent discovery of a new and unexpected type of meson (remember all the other mesons could be accounted for from the u, d and s quarks). One group named the particle the J and the other the ψ. Many people still refer to it as the J/ψ. A ready-made explanation existed for this new particle—it was the lowest mass version of the $c\bar{c}$. Acceptance was not immediate, but it was the only explanation that did not have serious drawbacks. Any lingering doubts about the existence of the charm quark were dispelled during 1975 and 1976 when more particles containing the charm quark were discovered. In 1976 Samuel Ting and Burton Richter (the leaders of the two groups) were awarded the Nobel Prize for their joint discovery.

As one might imagine, the story did not end there.

A new lepton, the tau, was discovered in 1975 (for this discovery Marty Perl received a share of the 1995 the Nobel Prize) that once again destroyed the symmetry between the number of quarks and leptons. Physicists immediately went on alert to look for a new quark. On 30 June 1977 Leon Lederman announced the discovery at Fermilab of the upsilon meson that turned out to be the $b\bar{b}$.

Evidence for the tau-neutrino is indirect but convincing and if nature requires equal numbers of quarks and leptons then the top quark must exist. In 1995 Fermilab announced that a two-year run of experiments produced evidence for the top quark. Top had been found after an eighteen-year wait.

The argument for equal numbers of quarks and leptons is not simply that of elegance ('if I were God, that is how I would do it!') but is also required by some theories that go beyond the standard model. However, the reader is entitled to ask how we can be sure that the progression stops at the third generation. Recent evidence produced at CERN has enabled us to 'count' the number of generations and establish that there are only three. We shall refer to this again in chapter 11.

9.5 Summary of chapter 9

• The existence of quarks was originally a theoretical idea that brought order to the pattern of hadron properties;

• the deep inelastic scattering experiments can be interpreted as evidence for high speed electrons interacting with localized charges within protons and neutrons;

• jets can be explained by the materialization of quarks and antiquarks out of the electromagnetic energy produced by e^+e^- annihilations;

• the R parameter is a means of counting the number of quark flavours that can be produced in e^+e^- annihilations as the energy of the reaction is increased;

• the R parameter provides evidence for the existence of a new quark property known as *color*;

• color plays the role of 'charge' in the strong force that electrical charge plays in the electromagnetic force;

• the discovery of the J/ψ meson was evidence for the charm quark that had been suggested in order to balance the numbers of quarks and leptons;

• more recent discoveries have taken us to the current state of six confirmed leptons and six quarks;

• there are some sound theoretical reasons for believing that there must be equal numbers of quarks and leptons.

Notes

[1] Gell-Mann named them quarks, Zweig coined the term aces. Quarks won, perhaps because there are four aces in a pack of cards—in which case the name was slightly prophetic.

[2] Later to become famous as the inventor of the Geiger counter.

[3] In 1914 Rutherford described it, perhaps uncharitably, as being 'not worth a damn'.

[4] The discovery of quantum mechanics, the physics of motion that must be applied within atoms, has shown that electron 'orbits' are far more complicated than planetary orbits.

[5] The effect is increased by the relativistic time dilation. A fast moving object will always see other particles to be moving more slowly as external time slows down when you are moving close to the speed of light.

[6] I will follow the American spelling to emphasize that this is *not* the colour that we see with our eyes.

Chapter 10

Experimental techniques

In this chapter we will look at various devices that are used to accelerate and detect particles. Modern accelerators are reaching the practicable limit of what can be achieved with the techniques that are in use. It is becoming clear that if we wish to work at energies very much higher than those that are currently available, then some new technique for accelerating the particles must be found.

10.1 Basic ideas

When a particle physics experiment is performed the procedure can be split into three stages:

1. *Preparing the interacting particles*
   This may involve producing the particles, if they are not common objects like protons or electrons; accelerating them to the right energies and steering them in the right direction.

2. *Forcing the particles to interact*
   This involves bringing many particles within range of each other so that interactions can take place. This must happen frequently so that large amounts of time are not wasted waiting for interesting interactions to take place.

3. *Detecting and measuring the products*
   The particles produced by the interaction must be identified and their energies and momenta measured if the interaction (which is not directly observed) is to be reconstructed.

These three stages are, to a degree, independent of each other. Each poses its problems and has forced technology to be improved in specific ways so that modern particle physics experiments can work.

10.2 Accelerators

10.2.1 Lawrence's cyclotron

The first real particle accelerator was developed by Ernest Orlando Lawrence, an associate professor at Berkeley, between 1928 and 1931. His *cyclotron* was the first device to combine the use of an electrical field to accelerate the charged particles with a magnetic field to bend their paths.

Figure 10.1 shows a diagram of the first cyclotron built to Lawrence's design. The device consists of two hollow 'D' shaped pieces of metal placed between the circular poles of an electromagnet. A small radioactive source placed at the centre of the device emits charged particles. The paths of these particles are deflected by the magnetic field (see page 21) and they move into circular arcs passing from inside one D into the other. While inside the volume of a D the charges experience no electrical force, even though the metal is connected to a high voltage. However, as the particles cross the gap between them they move from the voltage of one D to the different voltage of the other. This change in voltage accelerates the particles as they cross the gap.

To make sure that the particles are accelerated rather than slowed down by the change in voltage, the D that they are moving into has to be negative with respect to the D they are emerging from (if the particle is positive).

The particle now moves along a circular path through the second D, but at a slightly higher speed. When it crosses back into the first D at the other side of the circle, it experiences a change in voltage again. However, this time the voltage would be the wrong way round to accelerate the particle and it would be slowed down. The trick is to arrange to swap the sense of the voltage round while the particle is

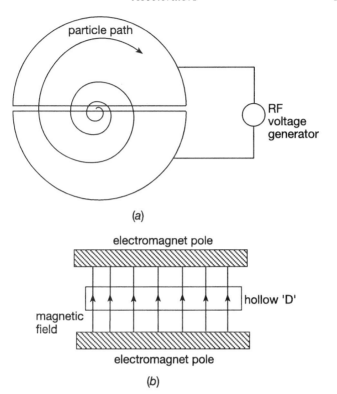

Figure 10.1 Lawrence's cyclotron: (a) plan view, (b) cross section.

inside the second D. Consequently, by the time it arrives at the gap, the voltage is always in the right direction to accelerate the particle and it receives a small kick every time it crosses a gap (twice per revolution).

Organizing the voltage to swap over at the right time would be difficult if it were not for a curious property of the particle's motion. Every time the particle is accelerated, you would think that it arrived at the gap a little faster than last time. However, as the speed of the particle increases so does the radius of the circular path. This means that a faster particle is deflected into a bigger circle and so has further to travel. Even though it is moving faster, it still arrives at the gap at the same time. Hence the changing voltage can be switched at a constant rate and it will always match the motion of the particle.

length of path $= 2\pi r$

where r is the radius of the circle; the time taken is

$$T = \frac{2\pi r}{v}$$

however, $$r = \frac{mv}{Bq}$$ (page 23)

$$\therefore \quad T = \frac{2\pi mv}{Bqv}$$

$$= \frac{2\pi m}{Bq}$$

so T is independent of the speed at which the particle is moving. The frequency with which the field should be switched over is

$$f = \frac{1}{T}$$

$$= \frac{Bq}{2\pi m}$$

f has to be at radio frequencies (RF) for a typical sub-atomic particle.

The first cyclotron measured 13 cm in diameter and accelerated protons to 80 keV energy. The device could be made bigger to achieve more energy, but eventually two fundamental limitations were met. Firstly, the circular magnets were becoming impracticably large, and secondly the particles were being accelerated to relativistic speeds and the constant time to complete a revolution no longer applied. This places a limit of about 30 MeV on a proton cyclotron.

The first development was to alter the frequency of the RF field to keep in step with the accelerating particles. This means that the accelerator can no longer work with a continuous stream of particles. Particles at a later stage of acceleration would be moving faster than those that had just entered the accelerator, and so it would be impossible to keep the field in step with all of them at the same time.

The answer was to accelerate particles in bunches to the edge of the machine, then extract them before a new bunch entered the accelerator.

While the bunch was being accelerated the radio frequency (RF) was adjusted to synchronize with them crossing the accelerating gap. The device was called a *synchrocyclotron*. Modern accelerators still accelerate particles in bunches. In 1946 a synchrocyclotron accelerated deuterium nuclei to 195 MeV.

Although the synchrocyclotron overcame the problem of the relativistic limit, it did not address the issue of the magnets.

As the demand for higher energies continued it became harder to increase the strength and size of the magnets. The next step was to use a series of magnets round an accelerating ring rather than two large magnetic poles.

In this design of machine the particles are kept on the *same* circular path by increasing the strength of the magnetic field as they become faster. The accelerating RF fields are applied at several points along the ring and the frequencies varied to keep in step with the particles. Such a machine is called a *synchrotron* and is the basic design followed by modern particle accelerators throughout the world.

10.2.2   A modern synchrotron

Figure 10.2 shows the most advanced form of particle accelerator now in use. A variety of different magnets bend and focus the beam with radio frequency cavities to accelerate it.

Quadrupole magnets

The quadrupole magnets (figure 10.3) are designed to kick particles that are drifting sideways out of the bunch back into line. They can only do this in one plane and they tend to move particles out of line slightly

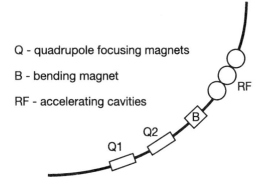

Figure 10.2   Part of a modern synchrotron.

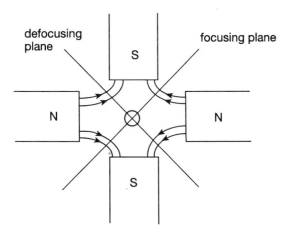

Figure 10.3   A quadrupole magnet. The beam runs perpendicular to the page through the circle at the centre of the diagram.

in the perpendicular plane. To overcome this they are used in vertical and horizontal pairs.

Bending magnets

Bending magnets are often made with a superconducting coil to allow large magnetic field to be maintained without expensive electricity bills. These magnets have their fields steadily increased as a bunch is accelerated to keep the particles in the same circular path.

RF cavities

The RF cavities are designed to accelerate the particles as they pass from one cavity into the next. They are designed to switch their potentials at radio frequencies keeping the bunch of particles accelerated as they pass through the system. One of the largest accelerating rings in the world is the Large Electron Positron Collider (LEP) at CERN in Geneva. LEP uses 1312 focusing magnets and 3304 bending magnets round its 27 km circumference and accelerates e^+ and e^- to an energy of 50 GeV each. The e^- and e^+ are sent in opposite directions round the same ring and collided at several points where detectors are placed.

Although the basic synchrotron ring structure dominates current particle accelerator design it does have some disadvantages.

Disadvantages of synchrotrons

- Bending magnets are reaching the limits of their field strength, so to go to higher energies very large rings will have to be constructed. LEP's 27 km is nearing the practical limit for such a ring—in its construction new surveying techniques had to be developed. Superconducting magnets are capable of much higher field strengths. They require a great deal of 'plumbing' to supply them with the coolant necessary to keep the magnets at superconducting temperatures. Again there is a practical limit to the size of the ring that can be supplied with superconducting magnets.

- All circular accelerators suffer from synchrotron radiation. Any charged particle moving in a circular path will radiate electromagnetic radiation. The intensity of this radiation depends inversely on the fourth power of the particle's mass ($\sim 1/m^4$) and inversely on the radius of the path ($\sim 1/r$). Consequently electron rings produce much more radiation than proton rings and large diameters are to be preferred. The energy lost by particles in producing this radiation had to be replaced by accelerating the particles again[1].

10.2.3 Linacs

A linear accelerator (linac) in which the particles move along a straight line through a series of accelerating cavities, does not suffer from synchrotron radiation. This makes them very useful in providing a preliminary acceleration for particles before they are inserted into a big ring.

However, there is one existing linear accelerator that is in use in its own right. The Stanford Linear Accelerator Centre (SLAC) uses a 3 km long linear accelerator that has recently been upgraded to accelerate both e^- and e^+ to 50 GeV along the same line and then to bend them round at the end to collide with each other. In its previous guise the SLAC linac was used to accelerate e^- to 20 GeV for use in the deep inelastic scattering experiments.

The obvious disadvantage of linacs (their length if they are to achieve high energies) places a severe limit on how much more they could be developed.

10.2.4 List of current accelerators

- *CERN: European Particle Physics Laboratory, Geneva*
  (i) LEP: collided e^+ with e^- at 100 GeV now upgraded to 180 GeV, 27 km tunnel
  (ii) LHC: planned to collide p and $\bar{p}$ at 14000 GeV using the LEP tunnel
- *FERMILAB: Chicago*
  Tevatron: collides p and $\bar{p}$ at 1800 GeV
- *HERA: Hamburg, Germany*
  collides 820 GeV p with 26.7 GeV e^-
- *SLAC*
  (i) SLC: collides e^+ with e^- at 100 GeV at the end of the 3 km linac
  (ii) PEP: e^+e^- ring colliding at 30 GeV—discontinued now; PEP II will be commissioned in 1999 and will be tuned to produce B quarks.

10.3   Targets

There are two ways in which particles are forced to interact with each other: by using a fixed target or colliding them with each other. Each has its own specific advantages and disadvantages.

10.3.1   Fixed targets

In a fixed target experiment, one of the interacting particles is contained in a block of material which acts as a target into which a beam composed of the other particles is fired. The target particle is assumed to be at rest while the accelerated energy and momentum are in the beam particle. Targets can be blocks of metal, tanks of liquid hydrogen that form a proton target (called bubble chambers) or silicon vertex detectors (see later in section 10.4). The latter two fixed targets have the advantage of being able to reproduce particle tracks as well as acting as providing particles for the reaction.

Advantages

- Easy to produce many interactions as the beam particles have a high probability of hitting something as they pass through the target;
- by using a bubble chamber or vertex detector it is possible to see the point in the target at which the interaction took place;
- fixed targets provide a convenient way of producing rare particles that can themselves be formed into a beam (e.g. kaons, antiprotons, pions etc).

Disadvantages

- As the beam particle carries momentum and the target particle has none, there is a net amount of momentum that must be conserved in the reaction. This means that the produced particles must carry this momentum. If they have momentum, then they must also have kinetic energy. So not all the energy in the reaction can be used to create new particles. This sort of experiment is not an efficient use of the reaction energy.

10.3.2 Colliders

In this type of experiment, two beams of accelerated particles are steered towards one another and made to cross at a specified point. The beams pass through each other and the particles interact.

Advantages

- It is easier to achieve higher energies as both particles can be accelerated rather than having to rely on just the one to carry all the energy;
- if the beams collide head on and the particles carry the same momentum, then the total momentum in the interaction is zero. This means that all the produced particles must have a total of zero momentum as well. This is a much more efficient use of energy as it is possible for all the reaction energy to go into the mass of new particles (if they are produced at rest)—no kinetic energy is needed after the reaction.

Disadvantages

- It is difficult to recreate the actual point at which the particles interacted—which is necessary if momentum and direction are to be measured and short lived particles (which may not travel very far from that point before they decay) are to be seen;
- it is difficult to ensure that the particles hit each other often enough to produce an adequate supply of interactions. This is measured by the *luminosity* of the experiment—the number of reactions per second per cm^2 of the beam areas. Modern experiments have achieved improved luminosity over the earliest colliders (the greatest luminosity achieved so far has been at Fermilab: 1.7×10^{31} s^{-1} cm^2). This improvement and the more efficient use of energy over fixed target experiments makes the collider the experiment of choice in modern particle physics.

10.4 Detectors

All particle detectors rely on the process of ionization. When a charged particle passes through matter it will tear electrons away from the atoms that it passes. This will result in a free electron and a positive ion in the material. The number of ions formed by the particle is a measure of its ionizing power and this, in turn, depends on the charge and velocity of the ionizing particle. Clearly a neutral particle has an extremely small chance of ionizing atoms and is consequently very difficult to detect.

The way in which the ions formed are used to track the progress of particles depends on the detector that is being used. Although there are many detectors that have been specifically designed to operate in a certain experiment, there are some general purpose devices that are in common usage.

10.4.1 The bubble chamber

These devices have gone out of fashion now partly as they are not suitable for use in a collider experiment, but also because they cannot cope with the number of reactions per second required in modern experiments, which tend to deal with very rare reactions.

However, bubble chambers used to be extremely important. Their considerable advantage was that they acted as a target as well as a particle tracker.

A bubble chamber consists of a tank of liquid hydrogen that has been cooled to a temperature at which the hydrogen would normally boil. The liquid is prevented from boiling as the tank is placed under pressure. In this state the beam of particles is allowed into the tank and after a short time (to allow an interaction to take place) the pressure is released. Charged particles that have crossed the chamber will have left a trail of ionization behind them. As the pressure has been released local boiling will take place within the tank. The bubbles of gas will tend to form along the ion trail as the charged ions act as nucleation centres about which the bubbles form. These bubbles are allowed to grow for a certain time, and then a photograph is taken of the chamber that provides a record of the interactions. There is a compromise in how long the bubbles are allowed to grow before the photograph is taken. A short time means small bubbles, which are hard to see, but which allow precision measurements to be taken. A long time gives larger, clearer bubbles but the extra space that they take up means that it can be difficult to separate tracks that run close to each other. A typical bubble size would be a few microns.

Advantages

- The actual interaction point is visible;
- if the chamber is placed in a magnetic field the momentum of the particles can be measured by the bending of the tracks;
- the thickness of the tracks is related to the amount of ionization which can be an indication of the type of particle that made the tracks;
- the existence of neutral particles can be deduced if the particle decayed into charged particles within the chamber (see figure 10.4)
- can be very accurate (a few microns on a track).

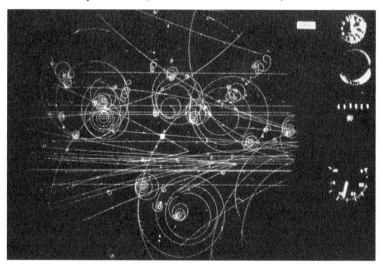

Figure 10.4 A typical bubble chamber picture. (Courtesy of the CERN press office.)

Disadvantages

- The chamber has to be re-set by compressing the liquid which bursts the bubbles. This means that there is a 'down time' during which any interactions that take place are not recorded. A very fast bubble chamber can go through 30 expansion cycles in a second, but this is nowhere near the rate at which events can be recorded with modern electronic detectors;

- does not produce instant information about tracks;
- needs a great deal of extra equipment to control the temperature, take the photographs and regulate the pressure.

10.4.2 Scintillation counter

A charged particle passing through a scintillating material will cause the material to emit a pulse of light. This light is produced by ionization of electrons which then recombine with an ion, excitation of electrons within the atom or even the breaking up of molecules (depending on the material). The light can be collected and detected using a sensitive photocell. This sort of chamber is used simply to detect the presence of particles and gives no information on the direction of travel.

10.4.3 Čerenkov detector

A Čerenkov detector is a more advanced form of scintillation counter. A particle passing through a material at a velocity greater than that at which light can travel through the material[2] emits light. This is similar to the production of a sonic boom when an aeroplane is travelling through the air faster than sound waves can move through the air. This light is emitted in a cone about the direction in which the particle is moving. The angle of the cone is a direct measure of the particle's velocity. A system of mirrors can ensure that only light emitted at a given angle can reach a detector, so that the Čerenkov chamber becomes a device for selecting certain velocities of particle. Alternatively, if the momentum of the particle is know (from magnetic bending) the Čerenkov's information on the particle's velocity enables the mass to be deduced so that the particle can be identified.

10.4.4 The multiwire proportional chamber (MWPC)

This detector has a precision nearly as good as that of a bubble chamber, but is able to work at a much faster rate as the information is recorded electronically.

A typical MWPC consists of three sets of wires in parallel planes close to each other (see figure 10.5). The outer two planes are connected to a negative voltage of typically −4 kV and the central plane is connected to earth (0 V). The whole collection is housed in a container that is filled with a low pressure gas.

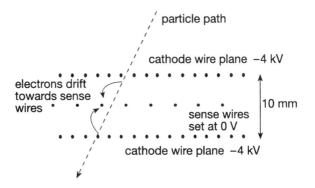

Figure 10.5 A typical MWPC cell.

The sense wires that form the central plane are spaced at 2 mm intervals and of a very small diameter (typically 20 μm). This has the effect of producing a very concentrated electrical field near the wire (figure 10.6).

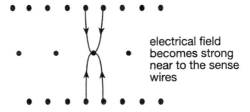

Figure 10.6 The field near a sense wire.

A charged particle passing through the MWPC will leave a trail of ionized gas particles behind. Under the influence of the electrical field the electrons from these ions will start to drift towards the nearest wire. As they get close to the wire they are rapidly accelerated by the strong field. The electrons are now moving fast enough to create ions when they collide with gas atoms. The electrons from these ions will also be accelerated and the process continues. Consequently a few initial ions can rapidly produce a substantial charge at one of the wires producing a measurable current pulse along the wire. This is known as the *avalanche effect*.

Electronic amplifiers connected to the sense wires can easily detect which wire produces the most current and so say which wire the particle passed closest to. They are also capable of providing information on more than one particle passing through the chamber at once.

An MWPC working in this fashion can produce a signal within 10^{-8} seconds of the particle passing through. By using a series of such detectors with the wire planes at angles to one another it is possible to reconstruct the path of the particle in three dimensions.

MWPCs have revolutionized particle physics experiments with their fast detecting times. Modern experiments produce huge numbers of interactions—far too many to be recorded on recording media such as computer tape. For this reason computers have to have rapid information about the nature of the interaction. MWPCs can pass information on to a computer within microseconds. The computers can then carry out a preliminary analysis of the event (typically taking a few microseconds) to decide if it is worth recording.

10.4.5   Drift chamber

Drift chambers measure the time taken for ions to arrive at the sense wires. A typical design of drift chamber is shown in figure 10.7. The central wire plane is set up with alternate field and sense wires and the voltages carefully chosen to produce a uniform drift speed for electrons along the chamber. This is a very delicate operation. The field strength between the wires has to be adjusted to compensate for the electron's loss of energy due to collisions with gas atoms within the chamber. If this is done properly the electrons will drift at a constant rate to the sense wire at which they will trigger an avalanche producing a detectable current pulse. As the drift speed is known, the drift time can be used to calculate the distance that the electrons had to move and so the distance of the particle's path from the sense wire is measured. With this sort of technique measurements are possible with an accuracy of a few tens of microns.

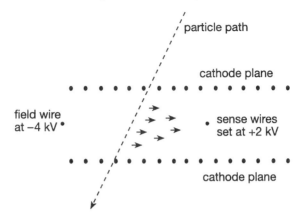

Figure 10.7 The drift chamber.

If the drift chamber is to work correctly another device must be used to start the clock running on the drift time when a particle enters the chamber. Typically this is a variety of scintillation counter.

10.4.6   Silicon detectors

The pressure of modern experimentation with very large numbers of events per second to be analysed and tracks recorded has led to the need for very high accuracy tracking equipment that can be read electronically.

Silicon detectors consist of strips of silicon (which is a semiconducting material as used in integrated circuits) which are very narrow (tens of microns) running parallel to each other. When a charged particle passes through the silicon it causes a current pulse in the strip that can be detected with suitable electronics.

The most up to date versions of these detectors are 'pixelated' which means that they are broken up into very small dots of silicon the way the photodetector in a camcorder is. Detecting which strip or pixel recorded a 'hit' and tracking these hits through a sequence of silicon detectors can give a very precise reconstruction of a particle's track. Micron accuracy is easily achieved. In addition, as these detectors are very fast and can be read directly by a computer they can provide information about an event that is virtually immediate, enabling the software that is controlling the experiment to decide if the event is worth permanently recording (on magnetic tape or optical disk) before the next event takes place.

10.4.7 Calorimeters

These are devices for recording the amount of energy carried by a charged particle. There are two types of calorimeter which work in different, but related, ways.

Electromagnetic calorimeters

These are designed to measure the energy carried by electrons, positrons or high energy photons. There are many different ways of constructing a calorimeter. One design consists of sandwiches of lead and a plastic scintillator. When an electron or positron passes through the lead it is frequently deflected by the atoms in the material. This can lead to the emission of high energy photons as the electron or positron loses energy in the process. The amount of lead is chosen to ensure that virtually all electrons or positrons in the experiment are brought to rest by radiating away all their energy in this manner. The high energy photons interact with the atoms of lead and this generally leads to their conversion into an electron–positron pair. These will, in turn, radiate more photons as they are brought to rest in the lead. In this way, an initial electron or positron can create a shower of such particles through the calorimeter. As this shower passes through the scintillator it causes flashes of light that can be detected by sensitive electronic equipment. How far the shower penetrates through the sequence of lead sheets and scintillator is directly related to the energy of the initial particle, as is the total number of particles produced in the shower. Photon energy can be measured as well, as high energy photons entering the lead will convert into electron–positron pairs and this will also set up a shower. The ability to detect and measure photon energy is often important in trying to pick up π^0's which decay into a pair of photons.

Hadron calorimeters

These are designed to measure the energy carried by hadrons. They also work by generating a shower from the original particle, but the process is not electromagnetic. One design uses iron sheets instead of the lead and the hadron reacts with an iron atom to produce more hadrons, which in turn react with other atoms. A shower of hadrons passes through the calorimeter setting off signals in the scintillator as they pass. The distance that the shower penetrates into the calorimeter gives the energy of the initial hadron.

10.5   A case study—DELPHI

The DELPHI detector is one of the four detectors specially designed and built for experiments at the LEP accelerator in CERN. DELPHI has been running since 1991. The first stage of the LEP accelerator collided electrons and positrons at 45 GeV per beam and so the total energy in the collision was exactly that required to produce a Z^0. Since then the beam energy has been increased (LEP2) to be enough to produce W^+ and W^- particles:

$$e^+ + e^- \;\rightarrow\; Z^0 \qquad\qquad \text{LEP1}$$
$$e^+ + e^- \;\rightarrow\; W^+ + W^- \qquad \text{LEP2}.$$

The purpose of LEP was to provide a tool that could probe our understanding of the standard model to very high precision. Between 1991 and 1994 LEP produced something like 4 million Z^0 events in the DELPHI detector.

Figure 10.8 shows the central part of the DELPHI detector. Other components fit on to either end to provide a complete coverage of the region round the collision point, which is at the centre of the detector. The whole detector, when assembled, is 10 m long and has a radius of 5 m.

The silicon detector is too small compared to the rest of the equipment to be shown on the diagram. It has a radius of 5.2 cm and is designed

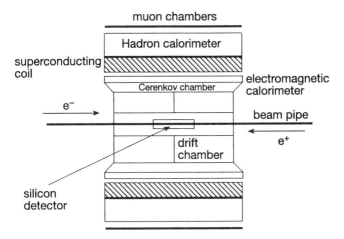

Figure 10.8 Part of the DELPHI detector.

to locate the interaction point and to feed information to the computer allowing the tracks coming from the interaction to be reconstructed.

The large drift chamber (3 m × 1.2 m radius) is divided into two halves vertically through the collision region. Ions produced by particles passing through the chamber drift towards the ends farthest from the interaction plane. The chamber provides accurate tracking data and, by measuring the amount of charge produced, some information helping to identify the particles passing through it. Particle identification is also provided by a complex Čerenkov chamber which is designed to help distinguish between protons, pions and kaons.

The electromagnetic calorimeter consists of layers of lead wires with gaps between them. An electromagnetic shower builds up as the electrons and positrons travel out from the centre of the detector through each layer of lead wires. In the gap between layers the shower ionizes a gas. The ions then drift along a layer under the influence of an electric field to sense wires at the end of the layer. Detecting the amount of charge arriving at the end of each layer charts the progress of the shower and so the energy of the initial particle.

The complete calorimeter is divided into rings along the length of the detector, each ring having 24 separate modules. This is to ensure that the chances of more than one shower being developed in a module at the same time are minimized.

The whole central part of the detector is in a magnetic field produced by the superconducting coil. This is the largest superconducting magnet in the world: 7.4 m long and 6.2 m in diameter. The field runs parallel to the beam pipe and the coil carries a current of 5000 A. To maintain its low resistance the coil has to be kept at a temperature of 4.5 K.

The final parts of the detector are the hadron calorimeter and the muon chambers. The iron in the hadron calorimeter is designed to stop all the hadrons produced in the interaction, so that anything passing into the muon chambers is bound to be a muon—the only particle that can get through the rest of the detector.

Figure 10.9 The drift chamber being moved into place in DELPHI. Note the layers of calorimeter in rings round the drift chamber. (Courtesy of the CERN press office.)

Figure 10.10 The DELPHI experimental hall. (Courtesy of the CERN press office.)

10.6 Summary of chapter 10

- The modern particle accelerator is a synchrotron;
- quadrupole magnets are used to focus the bunch of particles that are accelerated;
- bending magnets keep the particles in a circular path;
- radio frequency cavities accelerate the particles;
- circular ring accelerators suffer from synchrotron radiation (especially electron accelerators) requiring them to have a large diameter;
- linear accelerators do not produce synchrotron radiation, but have to be very long to accelerate particles to high energies;
- fixed target experiments produce lots of interactions, but are wasteful of energy due to the need to conserve momentum;
- collider experiments can be designed to have zero net momentum and so are much more energy efficient, but they are harder to produce lots of interactions with;

- there are a variety of detectors for identification (Čerenkov, drift chamber) and measurement (bubble chamber, MWPCs, drift chambers);
- calorimeters are designed to measure a particle's energy by the number of shower particles that it can produce when passing through dense matter.

Notes

[1] Even with a modest electron ring like that at DESY in Germany, which accelerates electrons to 20 GeV and has a radius of 256 m, 8 MW of electrical power are required to replace the energy lost due to synchrotron radiation alone.

[2] Before I get letters from the people who think that this violates relativity, I should explain that the particle must be moving faster than a light wave could if the light wave were to pass through the material. Light passing through matter never moves in an exactly straight line as it is scattered by the atoms. This means that a beam of light crossing a block of glass has to travel much further than the width of the block to reach the other side. We interpret this as the light moving more slowly in the block—hence refractive index. A charged particle moving through the material can get to the other side more quickly—hence it appears to move faster than light. The relativistic limit on velocity is the speed of light in a vacuum.

Interlude 1

CERN

CERN (the Conseil Européen pour la Recherche Nucléaire) is the largest research facility of its type in Europe and one of the largest in the world. The idea of establishing a major research facility that could be used by all European countries arose shortly after the end of the Second World War. The aim was partly political (it was seen as a step towards reconciliation by working together), partly to halt the drain of scientists away from Europe and partly in recognition of the increasing expense of advanced experimental science. No single country could afford to fund a worthwhile facility, but by combining their efforts financially and intellectually, Europe could play a significant role in fundamental research.

In June 1950 UNESCO adopted CERN as a central part of its scientific policy, the preparatory work having been done under the auspices of the European Centre for Culture. CERN was born on 1 July 1957 when the CERN convention was signed in Paris. In 1955 a 42 hectare site near the village of Meyrin close to Geneva was given over to the new facility. The site was extended by 39 hectares in 1965 by agreement with France and from that time on CERN has straddled the Swiss-French border (see figure I1.1). Another extension on French soil took place in 1973 when a further 90 hectares were made available.

CERN's main function is to provide accelerated beams of particles that can be used for experimental purposes. It develops and maintains particle accelerators and research facilities as well as the support staff required. Collaborations of physicists from various universities apply to run experiments on the site. CERN itself is funded by direct grant

Figure I1.1 A view of CERN showing the LEP ring and the SPS; the Swiss-French border is also shown. (Courtesy of the CERN press office.)

from the various member states. The experiments are funded by the universities from their research grants made available by government. Experiments can take many forms, from the small and cheap experiment that recently manufactured antiatoms of antihydrogen (PS210, see Interlude 2) to the very high cost and complex LEP experiments such as DELPHI. In the latter case an experiment can take more than 10 years from initial design through to construction and data taking followed by the careful analysis of the results. Such experiments are a considerable investment in manpower as well as finance. They are designed to be as flexible as possible. Many pieces of research are going on at once using the same equipment.

One of the reasons that CERN has been so successful is the careful way that it has developed its facilities. As our understanding has improved so the need to produce particles at higher and higher energies has driven the development. CERN has been able to use existing accelerators as part of the expansion. Nothing has been 'thrown away' if it could be made to work as part of a new machine.

Figure I1.2 shows the machines currently at work in CERN.

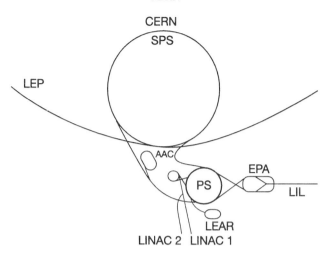

Figure I1.2 The accelerator complex at CERN.

- The PS, or Proton Synchrotron, was the first large accelerator to be built at CERN. It is 200 m in diameter and first started operation in 1959. The PS is able to accelerate particles up to 28 GeV. Nowadays it is used as a pre-accelerator for LEP and the SPS. It is fed with protons or ions made from heavy nuclei via two linear accelerators and a 50 m booster ring that takes their energy up to 1 GeV. After acceleration the protons can be passed on to the SPS. Alternatively the PS can take antiprotons from the Antiproton Accumulator (AAC) and inject them into the SPS.

- The SPS, or Super Proton Synchrotron, was commissioned in June 1976 to accelerate protons up to 400 GeV. It was later upgraded to 450 GeV. The ring is 2.2 km in diameter and runs underground at a depth of between 23 and 65 m. Between 1978 and 1981 the SPS was converted to enable it to accelerate protons and antiprotons simultaneously in opposite directions. These particles were then collided in the experimental halls. While working in this manner the SPS collider enabled the discovery of the W and Z particles in 1983.

- The Antiproton Accumulator (AAC) was designed to supply the SPS with antiprotons in its collider mode. Antiprotons have to be made in particle collisions and then stored until enough of them can be put into a bunch to be accelerated by the SPS. The AAC stores the antiprotons by circulating them in a 50 m ring.

- LIL is the linear injector for LEP. It is 100 m long and accelerates both electrons and positrons up to 0.6 GeV. The particles build up in the EPA (Electron–Positron Accumulator) and are then sent to the PS that accelerates them to 3.5 GeV. They are then taken up to 22 GeV by the SPS from which they are injected into LEP.

- LEP is 27 km in circumference and runs underground at a depth between 50 and 150 m. It started operation in July 1989 colliding electrons and positrons and has been recently upgraded to LEP2

Figure I1.3 A view of the LEP tunnel. The beam travels in a pipe inside the multiple magnets that run down the centre of the picture. To the left is the track for a monorail that runs the length of the tunnel. To the right are various servicing and cooling ducts. (Courtesy of the CERN press office.)

with a boost in energy. The Large Hadron Collider (LHC) will be built in the same tunnel as LEP.

- LEAR (Low Energy Antiproton Ring) is the only machine of its kind currently in the world. It takes antiprotons from the AAC and decelerates them to 0.1 GeV. It was recently used to supply antiprotons for an experiment to make antiatoms of antihydrogen.

CERN facts

Reference CERN annual report 1995.

Budget: 950 million Swiss Francs (1995)

Percentage contribution of the member states to the CERN annual budget:

Austria	2.77	Netherlands	4.77
Belgium	3.53	Norway	1.43
Czech Republic	0.10	Poland	0.12
Denmark	2.03	Portugal	1.13
Finland	1.04	Slovak Republic	0.05
France	18.25	Spain	5.18
Germany	23.17	Sweden	2.43
Greece	0.41	Switzerland	4.29
Hungary	0.14	United Kingdom	14.79
Italy	14.37		

Permanent CERN staff numbers:
105 research physicists
830 applied physicists and engineers
1065 technicians
430 craftsmen
508 office and admin. staff

Peak power consumption: 160 MW.

Chapter 11

Exchange forces

In this chapter we shall look in more detail at the general theory of forces. This theory has been extremely successful in providing a mathematical understanding of the four fundamental forces and has had a profound influence on the way that we think about the origin of the universe. The price has been that it is very difficult to extract from it a clear physical picture of what is happening. At the subatomic level forces are not simply 'pushes and pulls'.

11.1   The modern approach to forces

The modern generation of physicists have grown up with the idea of amplitudes, paths and all the mathematical machinery of quantum theory. Most accept that a complete physical picture of the subatomic world is impossible to achieve as we have no direct experience of this world on which to base our imaginations. Particle physicists have discovered that the notion of force has had to be radically altered in the light of quantum theory. The forces between particles are not the simple pushes and pulls that we learned about in school. As a result the intuition that we have developed in our experience of the physical world of cricket balls and bicycles is of no help to us. We must go back to basics and consider what actually goes on when we run a particle physics experiment. The fundamental basis of all particle experiments is illustrated in figure 11.1.

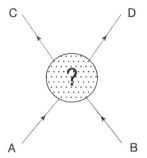

Figure 11.1 The bubble of ignorance.

Two particles, A and B, enter the experiment having been prepared in a specific way (known energy and momentum). They react and as a result two (or more) particles, C and D, emerge into detectors to be identified and their energy and momentum measured. In between preparation and detection a reaction happens. The nature and outcome of the reaction is determined by the fundamental force in operation and the energy in the reaction.

On the diagram, part of the reaction has been cornered off by a bubble. Inside the bubble the details of the reaction are hidden from direct view.

It is impossible to look directly at the reaction, as this would mean interacting with the particles *before* the reaction (in which case they would not be as we had prepared them) or *during* the reaction (in which case, we interfere with the reaction we are trying to study)[1]. For this reason we treat the region in the bubble as a region of ignorance and confine ourselves to trying to calculate the probabilities of different reactions (given a certain A and B, what are the chances of producing just those C and D's). Once we have calculated a full set of probabilities for all the possibilities, we can work something out about what might be going on inside the bubble.

The first set of reactions to have been fully explored in this manner where those due to the electromagnetic force. The resulting theory is called quantum electrodynamics (QED for short). For the fundamental work on this theory Feynman, Schwinger and Tomonaga were awarded the Nobel Prize in 1965. The theory is highly mathematical (and absolutely beautiful!), but we can gain some insight into its working by considering Feynman's approach to calculations which is built on the form of quantum mechanics that he developed.

11.1.1 Feynman diagrams

QED works by considering all the possible reactions that might take place inside the bubble. It calculates an amplitude for each possibility and adds them together to give the overall amplitude. This is a generalization of the sum over paths approach to calculating quantum mechanical amplitudes that we discussed in chapter 3.

Feynman developed a highly pictorial technique for making sure that all the possibilities were included. To illustrate this, we will consider a simple reaction:

$$e^- + e^- \rightarrow e^- + e^- \tag{11.1}$$

the elastic scattering of electrons. An obvious way in which this reaction may have taken place is illustrated in figure 11.2. The black dots on the diagram at the points where the photon is emitted and absorbed are called *vertices*. In figure 1.1 I illustrated such vertices as blobs, although in that case it was the weak force that was being discussed.

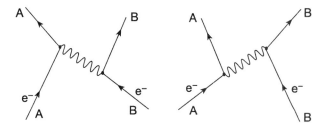

Figure 11.2 One-photon exchange.

If we wish to extract some physical picture of what is happening from these diagrams we imagine that as the electrons approached each other a disturbance was set up in their mutual electromagnetic field. The field is not shown on the diagrams, but the disturbance is represented as a photon linking the two particles. One diagram shows particle A emitting the photon that is then absorbed by B. The other diagram has the sequence the other way round. Our discussion of deep inelastic scattering suggests that the photon is formed due to the combined motion of *both* particles—so both diagrams represent what is happening. As there is no way we could experimentally tell them apart, *we have to count both.* For that reason we will abandon all attempts to tell the order in which things happen when they cannot be distinguished and draw a single diagram such as that in figure 11.3 (see pages 55–56).

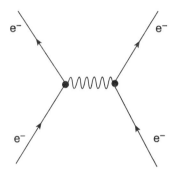

Figure 11.3 A Feynman diagram.

This is an example of a *Feynman diagram*; it is simply a sketch of a *process that includes both the time orderings in figure 11.2.* Feynman discovered a way of including both time sequences automatically in the same calculation. The diagram represents before and after (bottom and top of the diagram respectively) but all other time sense is lost.

This is not the only possible way in which the electrons might scatter off each other. Figure 11.4 shows some more Feynman diagrams that lead to the same result.

Each diagram represents a possibility that must be considered. Feynman derived precise mathematical rules that allow physicists to calculate the amplitude of each process directly from the diagrams. It is then a matter of adding the results in the correct manner to calculate the overall amplitude of the reaction.

Obviously there are a vast number of Feynman diagrams that can be drawn and would have to be included—apparently making it an

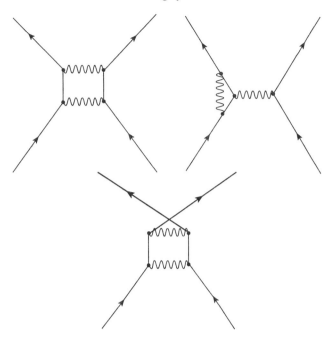

Figure 11.4 Some more complicated Feynman diagrams.

impossible task to calculate the final result. Fortunately this is not the case as QED is a 'well-behaved' theory. As the diagrams increase in complexity their amplitude decreases in size. To obtain a result accurate enough to compare with experiment only the first few diagrams need to be included. QED has been checked to 15 decimal places of accuracy on one of its predictions making it the most accurate theory that we have.

It is very natural to ask which of these multiple diagrams represents the truth—they cannot all be happening at the same time. This is essentially the same as asking which path the electron 'actually' travels along in the double slot experiment. Then we came to the conclusion that the quantum electron actually takes all paths at once. In this situation it is not a case of one of the diagrams being 'right' in different circumstances—they are all right every time the reaction happens! All we can say is that the disturbance in the electromagnetic field is very complicated and that each diagram represents an approximation to the actual physical process. Only by adding together all the diagrams can we get a mathematically correct answer[2].

Some authors only consider the first diagram in the sequence and develop the idea that the repulsion between the electrons is due to their exchanging a photon—as the diagram seems to suggest. Imagine two people cycling along next to each other. One throws a football towards the other. As a result he recoils and changes direction. The football then strikes the other cyclist causing him to be knocked away from his initial path. The photon is supposed to do a similar job to that done by the football.

While this is a quaint and easily visualized model, it does lead to severe problems if it is followed too far. For example, it becomes very hard to explain attraction.

11.1.2 The problem of attraction

Many people rapidly work out that the exchange football idea is inadequate to explain a force *pulling* particles together. However, when you draw the first Feynman diagram for the interaction between an e^+ and an e^- it looks exactly the same as figure 11.3 suggesting that the same process is at work.

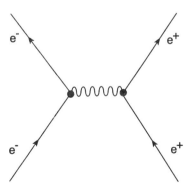

Figure 11.5 e^+e^- interaction.

Figure 11.5 may be the simplest Feynman diagram involved, but it is not the only one. To obtain a correct representation of the process *all* the relevant Feynman diagrams must be included. If we go to more complicated diagrams then we must include those that look like figure 11.4 as well (but with one of the electrons being an e^+). All the diagrams that we have drawn for e^-e^- reactions we can also draw for e^+e^-. However, there are some diagrams that can only be drawn for e^+e^-.

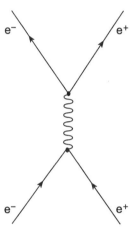

Figure 11.6 A distinct diagram for e^+e^- reactions.

Only when all the diagrams like figure 11.6 are included as well will the correct amplitude be obtained. The football exchange model is at best a glimmer of the true situation in one distinct case. It is better to surrender the need for a clear physical picture than to cause severe misrepresentations.

11.2   Extending the idea

Such is the success of QED that theorists have copied the technique to try and understand the other forces. This programme has met with considerable success over the last 20 years. We now have fully functioning theories of the strong and weak forces along the same lines as QED.

11.2.1   QCD

Quantum chromodynamics (QCD) is the theory of the strong force. In outline it is the same as QED. If we wanted to calculate the probability of two quarks interacting we would start by drawing a Feynman diagram such as figure 11.7.

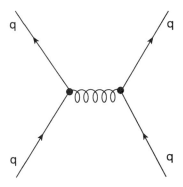

Figure 11.7   One gluon exchange.

The differences in the theory emerge when we have to use more complicated diagrams. We discovered in chapter 9 that quarks come in three different 'charges' or colors. The whole theory of QCD is built around this idea (hence the name of the theory). One of the implications is that the gluons themselves must be colored (but note that photons are *not* electrically charged in QED). This means that gluons also feel the strong force. Hence we have to consider some strange looking diagrams, such as figure 11.8.

Including such diagrams has a very interesting effect. The mathematical consequence is that as the diagrams become more complicated they become *more* important. Contrast this with QED in which the diagrams became less important to the total amplitude as they increased in complexity. QCD is not a 'well-behaved' theory and is very difficult to use in practice.

Including diagrams such as figure 11.8 also implies that the force between quarks *increases* with distance. This odd property of the strong force prevents us from observing isolated quarks in nature today[3].

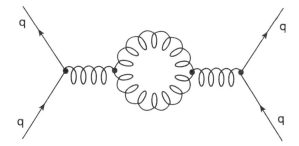

Figure 11.8 Gluon–gluon loops.

As the quarks get closer together the force between them gets smaller. Hence as our experiments increase in energy the quarks inside hadrons will get closer to each other when they interact and so we will be able to include fewer diagrams. Our approximations get better as our experiments increase in energy.

However, some of the most important questions, such as the masses of the hadrons (see chapter 6), cannot be completely answered at present. Some relatively new computer techniques[4] are making progress in this area.

11.2.2 Electroweak unification

One of the greatest recent successes of particle physics has been the development of a theory of the weak force. As an unexpected benefit the weak theory was found to be so similar to QED it became possible to develop a single mathematical theory that covered both forces. The complete theory is known as the *electroweak* theory. For this work Glashow, Salam and Weinberg received the Nobel Prize in 1979.

Merging the electromagnetic and weak forces is an ambitious project. The two forces seem quite distinct from each other. The electromagnetic force is effectively infinite in range, whereas the weak force is confined to distances less than 10^{-18} m. The electromagnetic force only acts between charged objects, but the weak force can act between neutral objects (neutrinos react with quarks). Furthermore the weak force changes the nature of the particles!

If one tries to write down a mathematical theory of the weak force along similar lines to QED it turns out to be impossible unless you allow there to be *three* types of disturbance of the weak field. In chapter 8 we saw that the weak field has two charged disturbances called the W^+ and the W^-; now this theory is telling us that in addition there is a neutral Z^0 disturbance.

In the first versions of this theory the W's and the Z were considered to be massless (just like the photon and the gluon) and so the Z seemed very similar to the photon. However, the simple theory failed as it could not explain why the weak force has such a short range.

The strong force is short range because the field energy increases as you separate the quarks (a consequence of the gluons carrying color) and the field decomposes into quarks when it gets too long (see chapter 7, page 131). This is not the case for the weak field. The theory suggests that the weak field behaves in the same way as the electromagnetic field, which has no limit on its range. The short range of the weak force must have a different explanation.

This problem held up the development of the weak theory until Higgs suggested a solution in 1966. The price of the solution is to introduce another field into physics—the Higgs field. The Higgs field does not cause a force in the way that the other four fields do; it interacts with the weak field. As W and Z particles pass through space they interact with the Higgs field and an exchange of energy takes place—the result of this being that the weak disturbances take on mass!

The whole subject of the mass of particles has been one of the outstanding problems of particle physics for 50 years. We discussed the problem of the mass of the proton in chapter 2. The Higgs field provides a way of explaining the masses of certain objects (W and Z for example), but at best it is only a partial solution as the values of the masses still cannot be calculated from first principles.

As a way of imagining how the Higgs field provides mass, consider an experiment in which a marble is fired from a gun inside a bath of treacle (the great thing about thought experiments is that you do not actually have to try to do them!). Given a certain force on the marble we should achieve an acceleration in air. In the treacle the acceleration will seem to be less due to the drag of the treacle. If we were ignorant of the physics of friction in liquids we might explain this result by saying that the mass of the marble was greater in the treacle than outside it.

The Higgs field acts like the treacle on the weak force disturbances. Unfortunately the Higgs field is always present but cannot be seen, so we can never do an experiment without it to see the 'actual' massless nature of the weak disturbances.

Steven Weinberg and Abdus Salam independently applied Higgs' idea to a combined theory of the weak force and the electromagnetic force. After the algebraic dust had settled a remarkable thing emerged. The equations contained five different fields—three that they had to introduce to make the weak force part of the equations work, one for the electromagnetic part and one Higgs field. However, the interaction between the Higgs field and the others caused the fields to mix up in an interesting way. Two of the three fields introduced by the weak force part of the equations took on mass—they are the W^+ and the W^-. The third field combined with the electromagnetic field in the theory. The resulting combination produced two particles—the Z^0 with mass and the photon which is massless. In other words the real particles that we observe are a mixture of the underlying fields brought about by the interaction with the Higgs field. The result is a rather complicated theory, but it has been tested by experiments and found to work in all instances.

The theory is also very specific about the way that the Z^0 interacts with other particles—it does not change the type of particle (in that way it is more like a photon than a W), but it can couple to objects with no charge (unlike the photon). This produces an effect known as 'neutral current' in which neutrinos can be seen to interact with other particles and stay as neutrinos. This type of reaction was discovered in 1973 and provided an early pointer to the theorists that they were heading in the right direction.

11.2.3 Experimental confirmation

In 1984 an experimental team at CERN lead by Carlo Rubbia announced the discovery of the W and Z particles. By colliding protons with antiprotons they had managed to muster enough energy to create the W and Z particles in sufficient numbers to be detected in an experiment. The masses were measured and found to agree with the theoretical predictions. Rubbia and Van Der Meer were given the Nobel Prize in 1984.

In 1990 the giant LEP accelerator started up at CERN. LEP collides e^+ and e^- at exactly the right energy to produce Z particles. These then decay in various distinct ways that can be seen in experimental detectors e.g.

$$e^+ + e^- \rightarrow Z^0 \rightarrow \mu^+ + \mu^-. \tag{11.2}$$

LEP is a Z factory. Millions of Z particles have been created and measured. The agreement with theory is very impressive.

Despite all this good news there is still one piece of evidence to be collected—the existence of the Higgs field has yet to be demonstrated. This is a very difficult thing to do as we know so little about it. We predict that it should be possible to create disturbances in the Higgs field—the so-called Higgs particles, but there is no evidence for them at present. They are a task for the next generation of particle accelerators.

11.2.4 Counting generations

One of the most important recent results to have emerged from the LEP accelerator is the experimental confirmation that there are only three generations of leptons and quarks. This generation counting has been done by measuring the lifetime of the Z^0. The Z^0 is capable of decaying into any pair of objects that have less mass than it does. This obviously includes all the neutrinos. In broad terms, the greater the number of different decays that a particle has available to it, the shorter its lifetime. The Z^0 lifetime has been measured and compared with calculations for two, three and four types of neutrino. The answer is clearly only consistent with three types of neutrino. From this we deduce that there are only three generations of lepton and by symmetry three quark generations as well. For the first time we can be sure that we have the complete list of elementary particles[5].

11.3   Exchange particles

This is quite a difficult section that can be missed out on first reading.

We have now built up an impressively consistent view of how forces operate between particles. In each case there is a force field and the interaction between the particles gives rise to the disturbance in the field. This disturbance carries energy and momentum and in a Feynman diagram is represented as a particle moving between the vertices. The electromagnetic disturbance is the photon, the strong force has the gluon and the weak force the W and Z particles. The collective name for these disturbances is *exchange particles*.

All the exchange particles are in principle massless, but the Higgs field interacting with the W's and Z's gives them very large masses. The mass of the W's is 86 GeV/c^2 and that of the Z^0 93 GeV/c^2, which represents an enormous amount of energy to have to muster in the weak field. Fortunately, we do not need all this energy to create the W and Z exchange particles. They can be emitted with very much less energy than this. When this happens they are said to be 'off mass shell', which is a very complicated sounding term but it just means that they have been emitted with less energy than they should have.

Now, an off mass shell particle is quite an odd thing. It certainly does not have the mass that we normally associate with the ordinary particle. It is a disturbance in the field that has been forced into existence rather than appearing naturally. In quantum mechanical terms, the diagrams that represent the emission of exchange particles off mass shell are similar to the non-classical paths followed by the electron in the double slot experiment. Quantum mechanics allows all possibilities to happen, but the further the particle is from being on mass shell, the smaller the amplitude that it will be emitted.

In mechanical terms, a reasonable analogy for the emission of a particle with less mass than it should by rights have is to compare it to the sound made when you lightly tap a wine glass. Every wine glass has its own natural tone, which it will produce when tapped. However, if we place a wine glass next to a loudspeaker and play a note very loudly, then the glass will try to vibrate with the same frequency as the tone being played. It will not be very successful and the vibration will die out rapidly after the tone stops (if you tap a glass the natural tone tends to last longer). We have forced the glass to vibrate at a frequency that is not natural. The price we pay is that the vibration does not last very long.

When two particles interact, the field between them is set into *forced vibration*. The result is the exchange particle that travels between them—but it is off mass shell. The closer the particles get to each other, the greater the energy in the field and the nearer to being on mass shell the exchange particle will be (the equivalent to the natural tone of the glass). This makes the production of the exchange particle much more likely.

This is why the weak force is short range. Trying to create W's and Z's without enough energy available for their full mass makes them difficult to create unless the interacting particles are very close (small amplitude). The photon and gluon do not suffer in this manner as they have no mass and so can be created relatively easily at any distance[6]. However, they can be off mass shell as well!

Consider colliding an electron and a positron. If they are moving with equal speed, then their combined momentum will be zero. On colliding they will annihilate as in figure 11.6. The photon in figure 11.6 must have zero momentum, or momentum would not be conserved in the reaction. This would mean that it was stationary. However, photons normally move at the speed of light. This is no ordinary photon! This one is off mass shell—this one has *got* mass. Such a photon is almost literally pregnant with energy and must convert back into a particle–antiparticle pair rapidly (as in figure 11.6).

Exchange particles are always off mass shell to some degree. This is why they can only be found between the vertices of a Feynman diagram. Any particles entering or leaving the bubble of ignorance must be on mass shell.

If you like, you can imagine that off mass shell particles do not exist at all. After all, we can never observe them. No one Feynman diagram represents what is going on inside the bubble of ignorance. They each provide an approximate way of looking at the complicated disturbance of the field. Only in total do they give an exact description of the interaction. We have found that the only way to make Feynman diagrams work as descriptions of interactions is to use off mass shell particles in the diagrams (like non-classical paths in the double slot experiment). Particles entering and leaving the bubble of ignorance are ordinary, on mass shell, objects. Some people refer to off mass shell particles as *virtual particles*, a name that emphasizes their nature— virtual reality is a simulation of reality, not the real thing.

11.4   Grand unification

Physicists are now rather flushed with success. They have produced a theory that seems to point to a basic link between two distinct forces. Furthermore there is a very suggestive piece of experimental evidence that points to further links. We mentioned earlier that the strong force decreases in strength when the energy of the particles increases. The strength of the electromagnetic force and the weak force also change with energy. The possibility is that they may all be the same strength at some energy. Figure 11.9 shows this variation qualitatively.

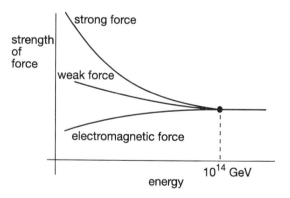

Figure 11.9 The variation of force with energy.

Note that the weak force appears to be stronger than the electromagnetic force on this diagram. This is because the graph is showing the 'inherent' strength of the force, which does not take into account the mass of the W and Z particles. As they are so massive, they are difficult to emit and this reduces the 'real' strength of the force to less than that of the electromagnetic.

All three forces become comparable in strength at the stunning energy of 10^{14} GeV. This is well beyond the range of any conceivable particle accelerator. However, this does not stop theorists from speculating about the nature of a 'Grand Unified Theory' (GUT) in which the strong, weak and electromagnetic forces are seen as the disturbances of a set of basic fields linked via a Higgs field. As a consequence there should be new fields that have not yet been seen with their own disturbances, the X and Y particles. The detailed properties of these particles depend on the exact version of the theory used, but their masses must be about 10^{14} GeV/c^2.

Although we do not know which GUT is correct at the moment, they all predict that protons should not be stable particles. The quarks inside should be able to emit X and Y particles and turn into leptons (hence baryon number is not truly conserved), but we are such a long way off the GUT energy that these processes must be very unlikely. As a consequence the lifetime of the proton is predicted to be more than 10^{30} years!

The greatest hope that the theorists have of testing their theories lies in speculating about the early universe. We believe that shortly after the big bang all the particles had energies above that required by the GUT theories—perhaps the nature of the interactions that went on then will provide some fingerprint on the universe that can still be seen today. We shall take up this idea in chapter 12.

11.5 Exotic theories

11.5.1 Spin

Spin is a property shared by all particles. In essence spin is a quantum mechanical property and there is nothing like it in classical physics. To a limited extent one can imagine that every type of particle spins on its axis, like a top, as it moves through space. The speed at which it spins is fixed for each particle. The units that we use to measure the rate of spin are in multiples of Planck's constant $h/2\pi$ (often abbreviated to $\hbar$). Quarks and leptons all spin at the same rate of $\frac{1}{2}\hbar$, photons, gluons and the other field disturbances have a spin of exactly $\hbar$. Even neutrinos spin, although the antineutrino spins in the opposite direction to the neutrino.

If the quarks have spin, then the hadrons made from them also must have some spin. A complex set of rules govern how the spins of the individual quarks combine to produce the spin of the hadron.

When they are in an atomic energy level electrons are allowed to spin on their axes as well as orbiting the nucleus. The direction of the spin can either be in the same sense as the orbital rotation or opposite to it (i.e. both clockwise or one clockwise and one anticlockwise)— the electrons are not allowed to 'roll' round in their orbits. A similar situation applies to quarks inside hadrons. When there are three quarks in a baryon the rules dictate that at least two quarks must be spinning in the same state. Consequently the total spin can be either $\frac{1}{2}\hbar$ (two in one sense and one opposite) in the case of the baryon octet, or $\frac{3}{2}\hbar$ (all quarks spinning in the same sense) in the case of the baryon decuplet. Protons and neutrons have $\frac{1}{2}\hbar$ spin, the Δ baryons are spin $\frac{3}{2}\hbar$.

In some mesons the quark and antiquark spin in the opposite sense so the overall spin of the composite particle is zero. Pions and kaons are spin-zero mesons. However, other combinations are possible as well: the J/ψ has a spin of $\hbar$.

Particles that spin with a 1/2 multiple of $\hbar$ are called *fermions* (i.e. leptons, quarks and baryons) those that spin with a whole number multiple of $\hbar$ are called *bosons* (i.e. mesons and the field disturbances). A promising modern theory attempts to link fermions and bosons and would allow a particle to convert from one type to another. Supersymmetry (SUSY) is a very elegant theory that is widely held to be true (at least in some respects) in particle physics circles. Unfortunately there is no experimental evidence for the particles that it predicts at present (although there have been some hints from the most modern results).

Supersymmetry was first discussed in the 1970s as an academic exercise. Theoreticians explored the mathematical structure of a theory that made no distinction between fermions and baryons, in the sense that the particles could be swapped round without altering the theory. They discovered that if the equations where to work they had to include a reference to a field that had not been originally specified in the theory. The properties of this field were strongly reminiscent of the gravitational field as described by Einstein's theory of general relativity. This exciting result held the first hint that it might be possible to draw gravity into the realm of particle physics.

However, there was a price to pay. The theory implied that there must be a fermion equivalent to every boson and vice-versa. The spin-1 equivalent to a quark is called a squark and that of a lepton a slepton. Along with this are the photinos (spin-1/2 photons) and the winos (spin-1/2 W's). Clearly this theory has allowed physicists' fluency with names to blossom. At a stroke the number of elementary particles has been doubled. Theorists counter this criticism by pointing out how neatly the theory manages to deal with the problems of a quantum gravitational theory. At the moment, the idea is viewed very positively. However, the supersymmetric particles have never been observed, implying that they are more massive than can be produced in our current accelerators. The theory requires some doctoring to make the supersymmetric particles have different masses to their common counterparts. The next generation of accelerators should settle the issue.

11.6   Final thoughts

If the principle of GUTs is correct, then we still have gravity to include in our scheme. At some time after the big bang the energies involved were such that gravity was comparable in size to the GUT forces. When this happens physics really does enter wonderland. The energy at which this is expected to happen is approximately 10^{40} GeV and all of our common-sense notions of space and time are rendered useless. This is an exciting area of physics. It is also the hardest to grasp mathematically and conceptually. Work in this area is far from complete, yet what seems clear is that when we have a complete theory that includes gravity as well, we will have totally revolutionized our understanding of space and time.

11.7 Summary of chapter 11

- Our understanding of forces comes from calculating the amplitudes for various particle reactions;
- we accept that we can never know exactly what happens in a given situation, but instead we have to consider all the possible interactions that give rise to the same result and add up their amplitudes;
- QED was the first theory to carry this programme out completely;
- Feynman's technique was to calculate the total amplitude of each interaction by listing all the possibilities in the form of diagrams;
- each diagram represents in a pictorial way a term in a mathematical approximation—it does not represent 'what happens';
- In QED the diagrams become less important as they become more complicated;
- QCD is the equivalent theory for the strong force;
- QCD is more difficult than QED as the diagrams become more important as they increase in complexity;
- the strong force gets weaker as energy increases, so it becomes possible to use approximation techniques at high energy;
- the weak force can only be dealt with by introducing a new field—the Higgs field;
- the Higgs field gives the W particles mass and combines the third weak particle with the basic electromagnetic field to produce the Z^0 and the photon—hence QED is combined with the weak theory;
- the W and Z particles have large masses, which makes the force weak and short range;
- GUT theories predict that the proton will decay and that B is not conserved;
- the only way of testing GUTs and more exotic theories is to see what they say about the early universe.

Notes

[1] What makes the difference here is that it is impossible to interact with a subatomic particle in a *small* way. We can only interact with them via other subatomic particles, and that takes place via the sort of reaction that we are trying to study!

[2] Compare this to the following situation. It is well known that π is approximately 22/7; it is actually the sum of the following sequence:

$$\pi = 4(1 - 1/3 + 1/5 - 1/7 + 1/9 - \cdots).$$

If you only use a few of these terms, then you will get an approximate value for π. In the same way, if you only use a few Feynman diagrams you get an approximate view of the interaction.

[3] However, as we shall discuss in chapter 12, in the early phases of the universe all the quarks were so close together that the forces between them were zero and they acted like separate objects.

[4] Parallel processing computers have been developed partly under pressure from particle physicists to help with QCD calculations.

[5] Perhaps I ought to stand back from that statement a little. If there are any neutrinos with masses greater than the Z^0 then this technique would not count them. This, however, seems very unlikely. It is possible that there might be smaller objects inside both quarks and leptons, but we have managed to count the constituents of matter at this level.

[6] The reason why the strong force is short range is that when enough energy is built up in the strong field a quark–antiquark pair tends to pop up rather than a gluon.

Interlude 2

Antihydrogen

Physicists believe that matter and antimatter behave in exactly the same manner in most circumstances. Certainly antimatter particles do not have 'negative mass' or antigravity properties and so they are not the answer to many propulsion problems as assumed in some of the worst science fiction.

However, it would be sensible science to check this and so there are many experiments going on at the moment to see if matter and antimatter particles have the same mass. Unfortunately, it is very difficult to experiment with antimatter. If an antimatter particle comes into contact with a the matter particle of the same type, then they will annihilate each other into energy. The amount of energy released in a single reaction of this sort is tiny and poses no safety risk—the problem is keeping the antimatter particles away from the matter particles for long enough to experiment on them (remember any equipment used will be made of the matter particles!).

At CERN they have partially solved this problem by building the Low Energy Antiproton Ring (LEAR). LEAR can store samples of antiprotons by circulating them back and forth in a 20 m diameter ring for several days. The antiprotons circulate in a small vacuum pipe and are kept in the centre (well away from the walls) by bending magnets. At several points round the ring there are detectors that monitor the position of the antiprotons. If they start to move out of place a signal is sent across the diameter of the ring to focusing magnets on the other side. By the time the antiprotons arrive the magnets are ready to squeeze them back into place. The developer of this technique, Simon Van der Meer, was awarded a share of the Nobel Prize in 1985.

In September of 1995 Professor Walter Oelert and his team from several European universities announced that they had used antiprotons from LEAR to successfully manufacture atoms of antihydrogen (experiment PS210). This was the first time that a complete anti-atom had been made. Antihydrogen consists of an antiproton with a positron in orbit round it. Physicists were quite sure that such objects should exist—indeed it should be possible to make anti-versions of all the chemical elements—but the experiment was a welcome confirmation of their ideas.

The experiment worked by allowing antiprotons from LEAR to pass through a small jet of xenon gas (see figure I2.1). Occasionally an antiproton passing close to a xenon nucleus would interact with its electromagnetic field and emit a photon. This photon might then convert into an electron–positron pair.

Although it is possible for any photon to do this, the electron and positron are normally a long way off mass shell as the photon does not contain enough energy to genuinely materialize this pair of particles. This would mean that the e^+ and e^- re-annihilate very soon after they are created. However, it is possible that during the short time in which they exist one of them could absorb a photon emitted from the xenon nucleus. If one of them were already close to being on mass shell, and the other, which was a long way off, was the one to absorb the photon, then it could provide the extra energy needed to supply its full mass. This would then mean that both the e^+ and the e^- could exist indefinitely.

The chain of circumstances needs to go one link further to create antihydrogen. The positron has to be moving slowly enough to be captured by the antiproton as it passes.

Clearly the whole sequence of events is rather unlikely (the probability has been estimated as 10^{-17}!), but it does happen. In the course of three weeks antiprotons circled the LEAR ring and passed through the xenon 3 million times a second. *Nine* antihydrogen atoms were produced.

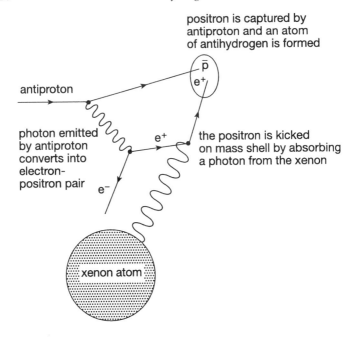

positron is captured by
antiproton and an atom
of antihydrogen is formed

p̄

e⁺

antiproton

photon emitted
by antiproton
converts into
electron-
positron pair

e⁻

e⁺

the positron is kicked
on mass shell by absorbing
a photon from the xenon

xenon atom

Figure I2.1 The chain of events leading to an atom of antihydrogen.

Once an atom of antihydrogen is created it is neutral and so is not affected by the bending magnets in the LEAR ring. The anti-atom flies off from the ring and can be detected with suitable equipment.

This experiment represents a first step in the study of anti-atoms. The antihydrogens produced only lasted for about 40 billionths of a second. This is far too short a time to enable anything other than a confirmation of their existence to be done. The next step is to try to trap some of these anti-atoms so that their properties can be studied in detail.

Chapter 12

The big bang

In this chapter we shall explore the evidence for the big
bang theory of the creation of the universe. We shall
also examine the implications of particle physics for this
theory and hence see why the physics of the microworld has
important implications for the universe as a whole.

12.1   Evidence

'It is a capital mistake to theorize before you have all the
evidence. It biases the judgement'   —   *Sherlock Holmes*
From *A Study in Scarlet* by Sir Arthur Conan Doyle.

Today, most astronomers and astrophysicists believe that the universe
was created in an extraordinary event, referred to as the big bang, some
15 billion years ago. This belief rests on three pieces of observational
evidence that fit neatly into the scenario of a big bang creation. Other
explanations for these observations have been offered, but none of them
have the simplicity or inter-relatedness of the big bang idea. We shall
look at each piece of evidence in turn and then examine how the big
bang accounts for them.

12.1.1   Red shift

In 1929 Edwin P Hubble published the results of a series of observations
made with the 100-inch diameter reflecting telescope at Mount Wilson,

near Los Angeles. Hubble had discovered that the spectrum of light from distant galaxies was shifted systematically towards the red wavelengths when compared with the light emitted from nearer galaxies. Furthermore, the amount of shift was directly proportional to the distance to the galaxy. This relationship has become known as Hubble's Law and its discovery marks the start of modern observational cosmology.

Stars and galaxies emit a wide range of wavelengths in the electromagnetic spectrum.

In the central cores of stars, the atoms are interacting with one another to such an extent that the normal line spectra of isolated atoms are merged into a continuous band[1]. This light has to pass through the outer layers of the star to reach us. The outer layers are less dense (hence the atoms act individually as in a gas) and much colder than the core so the atoms will be tending to absorb light rather than emitting it. Hence, the continuous spectrum of light produced in the core of the star is modified by having dark lines cut into it where atoms in the outer layers have absorbed some wavelengths.

By looking at the patterns of wavelengths that are absorbed, astronomers can tell which atoms are present in the star.

These dark lines, called Fraunhofer lines after the German optician who discovered them in the sun's spectrum[2], allow us to measure the extent to which a spectrum has been shifted. Ordinarily, if a spectrum is shifted towards the red wavelengths, then the longest wavelengths become longer, i.e. they become infra-red (IR) and hence invisible, but at the other end the ultra-violet (UV) wavelengths become violet and hence visible. The visible part of the spectrum is therefore unchanged (see figure 12.1).

The Fraunhofer lines are produced in unique wavelength patterns characteristic of each atom. If they are shifted, the astronomer can see by how much they have been moved along the spectrum (see figure 12.2). Without the Fraunhofer lines we would not be able to tell that a continuous spectrum had been shifted.

Given the wavelength, λ, of a line within the spectrum and the measured

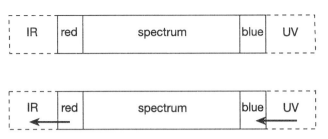

If the spectrum is shifted towards the red, then the UV light
becomes blue and the whole spectrum is unchanged

Figure 12.1 The shift of a continuous spectrum cannot be seen (the
dotted regions mark the invisible wavelengths).

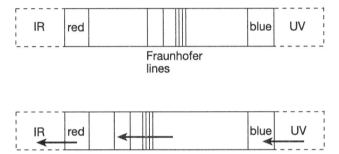

If the spectrum contains Fraunhofer lines, then
they will also move away from their correct wavelengths

Figure 12.2 The Fraunhofer lines provide a 'marker' by which shift can
be measured.

wavelength λ' of that line in the light from a galaxy, then we define
the red shift of the galaxy as being:

red shift, $$z = \frac{\lambda' - \lambda}{\lambda}.$$

Hubble's observations suggest that the size of the red shift is
proportional to the distance to the galaxy

> ## HUBBLE'S LAW
>
> $$z = Hd$$
>
> where H = Hubble's constant,
> d = distance to galaxy,

a relationship that is neat, but not at all self-evident. Modern astronomers are deeply suspicious of anything that suggests that our position in the universe is in any way special. Why should the red shift of distant galaxies be proportional to their distance from *our* galaxy? The big bang theory provides a satisfying answer to this question.

12.1.2   Helium

The universe as we observe it today consists of about 70–80% hydrogen and 20–30% helium. All the elements heavier than helium, like the carbon from which we are made, are trace impurities added to this basic mix. Any theory of the creation of the universe has to explain the observed abundance of the elements. In 1957 a famous paper by Burbidge, Fowler and Hoyle showed that the atoms heavier than helium could be built up by nuclear fusion taking place inside stars and during the violent supernova explosions that can destroy certain types of older star. Since then astrophysicists have become convinced that all the heavier elements have been made by the fusing of lighter elements inside stars[3]. However, the observed abundance of helium cannot be explained in this way.

There are two basic reasons for this. Firstly, the amount of helium that we can see inside stars (by looking at the Fraunhofer lines) seems to be independent of the age of the star. If the helium content of the universe has been built up over long periods, then one would expect the older stars to contain less helium than the younger ones, which have been made out of more recent material. No such variation is observed. Secondly, the process of forming helium out of hydrogen releases a great deal of energy (this, after all, is how a hydrogen bomb works). The release of energy required to build up 25% helium in the universe would make stars shine far more brightly than they do.

Hence we require some other mechanism to explain why 25% of the mass of the universe is helium. Furthermore this mechanism must have produced all this helium *before* the oldest stars were formed.

12.1.3 The microwave background

In 1963 Robert Wilson and Arno Penzias made an accidental discovery that revolutionized our thinking about cosmology. They had been using a radio antenna, built by Bell laboratories in America for satellite communication, to study radio emissions from space when they found a low intensity signal in the microwave region of the electromagnetic spectrum. The signal appeared to be coming equally from all directions in space. Normally a radio source in space will be localized to a particular part of the sky. For example, the planet Jupiter in our own solar system is a very strong source of radio waves due to the constant thunderstorms that are taking place in its atmosphere. Certain types of galaxy are very strong emitters of radio waves and their noise can be identified as coming from a particular part of the sky. The hiss that Penzias and Wilson had picked up was the same intensity in every direction and at any time of the day or night. For this reason they first suspected that it was interference produced within the antenna itself. However, after a careful study and inspection of the device (including the eviction of some pigeons that had taken to nesting within the antenna) they could find no explanation for the noise. Eventually they were forced to conclude that it was coming from outer space.

Since the initial discovery in the centimetre wavelength range, measurements of this background radio noise have been made at a variety of other wavelengths. The results of these measurements have shown that the intensity of the noise varies with wavelength in a characteristic way well known to physicists—a black body spectrum.

The earliest studies showed that all objects that are in equilibrium with the radiation that they produce emit a characteristic spectrum that only depends on the temperature of the object (figure 12.3). A perfectly black object would naturally be in equilibrium with its radiation (i.e. it would be emitting and absorbing energy at every wavelength equally), so the spectrum has been named black body radiation. (Many objects that look black to the eye are not perfectly black. Any object that looks black absorbs all visible light, but it may not absorb IR or UV radiation—a perfectly black object would absorb all wavelengths equally.)

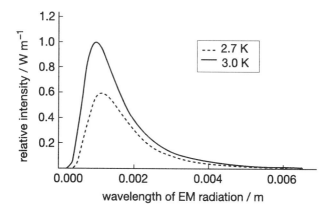

Figure 12.3 A black body spectrum.

As the temperature increases, so the wavelength at which the most electromagnetic radiation is produced decreases—the object glows red, then more yellow and finally white as the distribution of wavelengths produced falls within the visible spectrum.

Careful measurements have shown that Penzias and Wilson's radio radiation is equivalent to that which would be produced by a black body 'warmed' to 2.7 K. Furthermore, as the radiation is coming at us equally from all directions, the mechanism responsible must be widespread.

12.2 The expansion of the universe

The modern theoretical science of cosmology started in 1915 with the publication of Einstein's theory of general relativity. In this paper Einstein expanded the range of his earlier special theory of relativity to include gravity.

Einstein attempted to apply his new theory of gravity to the universe as a whole. To do this he imagined the matter of the universe to be smeared out into a uniform density. This is not as bad an approximation as you might think. Although there is some significant clumping of matter on local scales (stars, galaxies, etc) the distribution of clusters of galaxies across the universe appears to be quite uniform.

Having solved the complicated equations in some simple cases, Einstein was puzzled by the fact that his theory seemed to show that a universe could not be static—it must either be getting bigger or getting smaller. At the time the prevailing belief was that the universe was in a 'steady state' with no overall change taking place. Einstein chose to believe the opinion of the time rather than his equations and modified his theory to allow a motionless universe.

In 1922 Alexander Friedmann found a general solution to the original Einstein equations that showed categorically that Einstein's theory must produce a universe that is changing in size. The work was not widely discussed partly because of its highly mathematical nature and partly because the universe was still believed to be in a steady state. However, when Hubble published his data in 1925, Friedmann's results provided a natural explanation. Light from galaxies is red shifted because the universe is expanding.

Friedmann's work showed that space itself is not an empty volume through which galaxies are moving—space *itself* is changing. If we observe the distance between two galaxies to be increasing in a Friedmann universe, it is not because the galaxies are moving apart, it is because the space is swelling.

Most people are not used to thinking of space as a *thing*. To try to understand this idea, imagine the surface of an inflated balloon that is covered with ants. The ants will move about on the surface—this is like galaxies moving *through* an unchanging space. Now imagine that we put small piles of food on the surface of the balloon, so that the ants stay in place for an extended period. If we were to inflate the balloon further, the ants would all start to move apart as the surface of the balloon stretched. *But the ants would not be moving.*

It is the second case that is similar to Friedmann expansion—the ants are not moving *over* the surface, they are moving *with* it. The surface of the balloon represents the space in which galaxies are placed. Galaxies within the universe appear to be moving as the space they are contained in stretches. Furthermore, as light passes through a stretching universe its wavelength will also be stretched. This is why the light from galaxies is shifted towards the red end of the spectrum.

Clearly it is very difficult to visualize what is meant by space stretching. One has to try to abandon the old idea that space is nothing more than an emptiness through which something moves. When one initially thinks about the universe expanding the natural reaction is to imagine the galaxies getting further apart as they move into areas of the universe that originally did not have galaxies in them (like a gas introduced into the corner of a box diffusing throughout the whole volume). This is *not* what is happening. I have tried to illustrate the difference in figure 12.4.

In figure 12.4(b) the space has been expanded, but not the galaxies themselves. The expansion of the universe would cause the galaxies, stars and us to expand as well were it not for the local gravity and electromagnetism holding us together. The expansion of space can only be seen in the emptiness between galaxies and in the stretching of the light that crosses this region.

To see how this expansion gives rise to Hubble's law consider figure 12.5. In this diagram light is shown being emitted in our direction from two galaxies A and B. Each successive diagram is a larger magnification of the first one. The wavelength of the light is increased by that magnification as well as the distance to each galaxy.

By the time that a 50% expansion has taken place, the light from galaxy A has arrived at us red shifted by 50% . However, the light from galaxy B is still on the way to us as it started off further away. Another 50% expansion will have taken place by the time that light from B arrives. Hence B will be considerably more red shifted than A because the universe has expanded by a greater amount in the time that it takes to get to us.

Consider also what is happening from the point of view of galaxy A. An observer based in galaxy A would consider themselves to be stationary and all the other galaxies to be moving away from them. Friedmann expansion implies that *all* galaxies will observe the others to be moving away from them—there is nothing special about our galaxy.

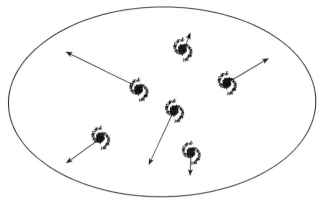

The universe as a box filling with galaxies

(a)

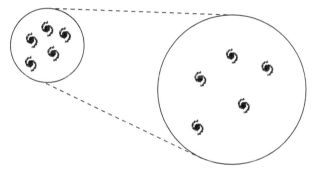

The universe as expanding space

(b)

Figure 12.4 (a) The natural idea is that galaxies are rushing apart filling previously empty regions of space. (b) The truth is that the universe is already full of galaxies; the space they occupy is getting bigger. In this diagram the bigger circle is a magnification of the smaller.

Hubble's law is a consequence of the expansion of the space through which light is travelling. However, it is often treated as an example of the Doppler effect, which is the shifting up in frequency of a sound produced by an object moving towards us, and down in frequency if the object is moving away. The same effect can be observed for any

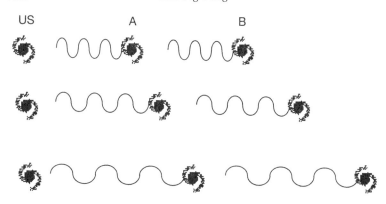

Figure 12.5 Hubble's law.

type of wave. It is possible to observe genuine Doppler effects in the light from stars, but the Hubble shift is not one of them.

Treating Hubble's law as if it were due to the Doppler effect is confusing. It is often explained by analogy with a genuine Doppler shift, such as that of the siren of a moving police car, and this gives the picture of galaxies moving *through* space. This is an incorrect picture—it is the space itself which is stretching and the wavelength of light is stretching with it. However, it is useful to calculate the *apparent* speed at which galaxies are moving away from us. This tells us about the rate at which the universe is expanding.

We can calculate this apparent speed by combining Hubble's law with the Doppler relationship:

$$z = u/c$$

where u = speed of light source, c = speed of light. Hubble's law states

$$z = Hd$$
$$\therefore \quad u = Hcd.$$

Hence the apparent speed of the galaxies is also proportional to their distance from us.

If the universe is currently expanding then logically there must have been a time when all the galaxies were together in one place. At some point all the stars would be right next to each other—in fact at this stage there would be no stars, just a super-hot plasma of hydrogen and helium gas. If we go back further, unwinding the expansion of the universe, then all the atoms must merge together. Protons and neutrons lose their individual nature and the universe becomes a sea of quarks, gluons and leptons. Particle physics becomes important. Further back in time still we start to go beyond what we currently understand. In principle the density of the universe continues to climb without limit as everything gets closer and closer together. This is the era of the big bang—the mysterious event that started the whole expansion going. The existence of this extraordinary starting point is a logical consequence of the observed expansion according to Friedmann's mathematics.

We can use the expansion law to estimate the time that has passed since the start of the expansion:

$$\text{velocity} = Hc \times \text{distance}.$$

If the velocity is constant, then the time it has taken a galaxy to reach its current distance is

$$\text{time} = \frac{\text{distance}}{\text{velocity}}$$

$$= \frac{\text{distance}}{Hc \times \text{distance}} = \frac{1}{Hc}$$

i.e. the age of the universe is $(Hc)^{-1}$. Measuring the value of H is a very complicated task and the final value is very uncertain[4]. Currently the best value is quoted as:

$$Hc \sim (15\text{--}30) \text{ km s}^{-1} \text{ (million light years)}^{-1}$$

i.e. that a galaxy 1 million light years away would be expanding away from us at between 15 and 30 km s^{-1}. Hence, if the expansion has been at a constant rate the time the galaxy has taken to get to 1 million light years away is:

$$\text{time} = \frac{1 \text{ million light years}}{(15\text{--}30) \text{ km s}^{-1}}$$

$$= \frac{9.45 \times 10^{18} \text{ km}}{(15\text{--}30) \text{ km s}^{-1}}$$

$$= (9\text{--}20) \text{ billion years}$$

hence the often quoted value of the age of the universe as being 15 billion years.

Friedmann's work provided a natural explanation for Hubble's law. Until that discovery, nobody took the idea of an expanding universe seriously, so Friedmann's paper languished (Einstein recommended that it not be published!). Now we are led to the logical conclusion that at some stage in the history of the universe, there must have been a period of incredibly high density. By exploring the physics of this period we can find natural explanations for the helium in the universe and the cosmic microwave background.

12.3   The physics of the first three minutes

12.3.1   The temperature of the universe

As we chart the history of the universe it is convenient to mark our progress in terms of temperature. The temperature of a gas is a measure of the average kinetic energy of the particles within it. We shall be dealing with particles that are moving at speeds near to that of light, so it is more appropriate to refer to their total energies. For a gas the temperature is related to kinetic energy by:

$$\langle KE \rangle = \tfrac{3}{2}kT$$
$$= 1.38 \times 10^{-23} \text{ J K}^{-1}$$

where T = absolute temperature in kelvin,
k = Boltzmann's constant.

The factor 3/2 varies depending on the type of particle involved, but it is generally ~ 1 so we shall be working on the assumption that in the universe the average energy:

$$\langle E \rangle = kT.$$

12.3.2   Thermal equilibrium

Gases are composed of particles moving with a wide variety of speeds. We are able to define a temperature for the gas because the particles are colliding with one another (and the walls of the container) in such a way as to keep the average energy constant (if some particles collide and speed up, some others collide and slow down—so the average remains the same). The gas is in *thermodynamic equilibrium.*

If the temperature of the universe as a whole is to mean anything, then the particles within it must be in thermodynamic equilibrium. However, the particles are not simply bouncing off one another, as happens in a gas; these collisions are particle interactions dictated by the fundamental forces. Particle physics helps us to understand the processes that kept the universe in equilibrium.

12.3.3   Threshold temperatures

For each particle species there is a threshold temperature. Once the universe drops below that temperature the species is effectively removed from the universe. As long as the temperature of the universe is above threshold, the following reactions can take place:

$$\text{particle} + \text{antiparticle} \rightarrow \gamma + \gamma \tag{12.1}$$

$$\gamma + \gamma \rightarrow \text{particle} + \text{antiparticle}. \tag{12.2}$$

These reactions are similar to the annihilations and materializations that we have been studying throughout this book. While these reactions are taking place the particle and its antiparticle will be active constituents of the universe. Obviously the photons involved have to have sufficient energy to provide the intrinsic energy of the particles, so that:

$$2E\gamma = 2mc^2$$

with m being the mass of the particle. Once the energy of the photons in the universe becomes too small for the materialization process to work (remember that the photons are always being cooled by the expansion of the universe—as they are shifted towards the red end of the spectrum their energy decreases) there will be no materializations to replace particles converted to photons by the annihilations, and that species is wiped out.

This will happen once the photon temperature drops below:

$$T = \frac{\langle E \rangle}{k}$$
$$= \frac{E\gamma}{k}$$
$$= \frac{mc^2}{k}.$$

As an example, consider the presence of protons in the universe. The threshold temperature of the proton is:

$$T = \frac{mc^2}{k}$$
$$= 1.1 \times 10^{13} \text{ K}$$

and so once the universe had cooled below this temperature the protons and antiprotons annihilated each other and the proton became a very rare object, compared to photons. Table 12.1 lists the threshold temperatures of various particles.

Table 12.1 Particle threshold temperatures.

Particle	Mass (MeV/c^2)	Threshold temperature ($\times 10^9$ K)
p	938	10888
π^+	140	1620
e^-	0.5	6
μ^-	106	1226

12.3.4 Radiation versus matter

As the universe expands the density of matter decreases (the same amount of matter occupies a larger volume). Cosmologists find it convenient to chart the expansion of the universe by the *scale parameter, S*. If the size of the universe could be measured at a defined point in time (today would be convenient) the universe's size at any other time could be calculated by multiplying every distance by the factor S for that time (if today was the reference, then for all times in the past $S < 1$). The volume of the universe varies with the scale factor S^3, so the matter density should vary as $1/S^3$.

The photons in the universe also contribute to the mass density via their energy. However, as the universe expands the photons are red shifted— their wavelength increases with the scale factor. The energy of a photon depends inversely on its wavelength, so the energy must vary as $1/S$. This is coupled with the expansion of the universe's volume, so the photon's contribution to the density of the universe must vary as $1/S^4$.

In the earliest phases of the universe the temperature was higher than threshold for all particle species. Hence matter and radiation were equally important in the universe. Furthermore all the particles were moving at speeds close to that of light, so in all cases[5]:

$$E = KE + mc^2$$

$$\text{but} \quad KE \gg mc^2$$

$$\therefore \quad E \sim KE.$$

Hence the matter behaved in a very similar manner to the radiation. This earliest phase of the universe is called the *radiation dominated* epoch. As the universe ages the densities of matter and radiation continue to drop, so eventually the density of matter becomes greater than that of radiation—the universe is now in the *matter dominated* epoch. This happened a few hundred thousand years after the big bang.

Detailed calculations using Friedmann's models show that the rate of expansion of the universe is different in the two epochs. One specific Friedmann model was promoted in a joint paper by Einstein and de Sitter[6]. It has the simplest dependence of scale on time:

radiation dominated epoch: $S \sim t^{1/2}$

matter dominated epoch: $S \sim t^{2/3}$

t being the time after the big bang.

12.3.5   The early history of the universe

We will start our account of the history of the universe at 10^{-43} seconds after the big bang. At this time the temperature of the universe was greater than 10^{33} K and gravity was as strong as the other three fundamental forces. This is the epoch of ignorance; our understanding of physical law fails us.

As the universe expanded and cooled towards 10^{33} K gravity decoupled from the other forces and the grand unified theories mentioned in chapter 11 apply. There are many competing theories and the physics of this epoch is far from certain.

In broad outline the universe is above the threshold temperatures of the X, $\bar{\text{X}}$, Y and $\bar{\text{Y}}$ particles that mediate the GUT force. Consequently the universe is populated with quarks, leptons and all the exchange particles. The numbers of quarks, antiquarks, leptons and antileptons are kept equal by the GUT interaction. However, as we reach the end of the GUT epoch the X and Y particles disappear from the universe. As they decay they leave behind a small excess of quarks over antiquarks. Detailed calculations suggest that as this epoch ends the universe should contain $10^9 + 1$ quarks for every 10^9 antiquarks[7]. We have now reached 10^{-35} seconds after the big bang.

From now until 10^{-12} seconds the universe's temperature is above threshold for the W and Z particles and so the electromagnetic and weak forces are equal in strength—the strong interaction has separated. Quarks and antiquarks are too close to one another in the universe for the strong force to bind them into hadrons. Any quarks that did temporarily bind together would be easily blasted apart again by collisions with the high energy photons present in the universe. This epoch is referred to as the *quark plasma*. At the end of this period the universe has expanded sufficiently for the quarks and antiquarks to bind into hadrons. In particular, protons and neutrons form. The tiny excess of q over $\bar{\text{q}}$ left over from the GUT epoch is now reflected in a tiny excess of p over $\bar{\text{p}}$ and n over $\bar{\text{n}}$.

At 10^{13} K and 7×10^{-7} seconds into history, the universe drops below threshold for protons and neutrons. As a result these particles stop being major constituents of the universe. There is a general annihilation of p with p̄ and n with n̄ , but the universe is left with the small excess of matter over antimatter. Before the annihilation the numbers of protons and neutrons were kept equal to those of the photons by the equilibrium reactions:

$$p + \bar{p} \rightarrow \gamma + \gamma \qquad (12.3)$$

$$\gamma + \gamma \rightarrow p + \bar{p} \qquad (12.4)$$

$$n + \bar{n} \rightarrow \gamma + \gamma \qquad (12.5)$$

$$\gamma + \gamma \rightarrow n + \bar{n}. \qquad (12.6)$$

After the annihilation we expect one proton to survive for every 10^9 photons in the universe. This is a number that can be measured today and is found to be comparable to the predictions from GUTs. The surviving protons and neutrons are still kept in thermal equilibrium with the rest of the universe via neutrino interactions:

$$\bar{\nu}_e + p \rightarrow n + e^+ \qquad (12.7)$$

$$\nu_e + n \rightarrow p + e^- \qquad (12.8)$$

$$e^- + p \rightarrow n + \nu_e \qquad (12.9)$$

$$e^+ + n \rightarrow p + \bar{\nu}_e \qquad (12.10)$$

which are recognizable as those that we met in chapter 4 when we discussed the solar neutrino problem. At this early stage of the universe the density of matter and the energies concerned are big enough for neutrino interactions to happen regularly. This keeps protons and neutrons in thermal equilibrium and also the number of protons equal to the number of neutrons. As the neutrino energy drops the mass difference between the proton and neutron becomes significant and reactions (12.7) and (12.9) happen less often than the others. Therefore there is a gradual conversion of neutrons into protons as the universe ages.

At about 10^{-5} seconds the universe drops below threshold for pions and muons and these particle cease to be major constituents of the universe.

At 1.09 seconds after the big bang the temperature has dropped to 10^{10} K and the density is such that the neutrinos can no longer interact with matter sufficiently often to maintain equilibrium. From this instant on they decouple from the universe. They do not disappear. They remain forever within the universe but do not interact with matter to any great extent. The expansion of the universe red shifts their energy just as it does for photons. They retain a black body energy spectrum and should presently have a temperature of approximately 2 K. Although we expect there to be ~500 neutrinos per cm^3 in the universe, we currently have no way of detecting them. Measuring the temperature of the neutrino background would be an important experimental confirmation of the big bang.

Once the neutrinos have decoupled, there is no process to keep the protons and neutrons in thermal equilibrium—reactions (12.7)–(12.10) stop. By this time the ratio of protons to neutrons (p/n) has shifted to 82% protons and 18% neutrons. From this time onwards, the decay of free neutrons will start to eat into their numbers further.

The next major milestone takes place three seconds after the big bang. This is the time at which the universe has cooled to 5.9×10^9 K and consequently the e^- and e^+ threshold has been reached. There is a net annihilation of e^+ and e^- leaving an excess of e^- due to the unequal numbers set up in the GUT epoch. The remaining e^- stay in thermal equilibrium with the photons as they are free charges and hence interact well with electromagnetic radiation.

3.2 minutes into the life of the universe the photon temperature drops to the point at which deuterium (pn nuclei) can form. Before this time any deuterium nuclei that formed would be blasted apart by photons. We have entered the epoch of nucleosynthesis in which helium is formed from deuterium (d) by nuclear reactions:

$$d + d \rightarrow {}^3He + n \qquad (12.11)$$

$$^3He + n \rightarrow {}^3H + p \qquad (12.12)$$

$$^3H + d \rightarrow {}^4He + n. \qquad (12.13)$$

The process stops at ^4He which is the most stable of all nuclei. As this 'cooking' of helium takes place all the free neutrons are swept up and bound into nuclei. Once bound in this way, the strong interaction between the protons and neutrons stabilizes the neutrons preventing them from decaying. By the time that nucleosynthesis starts, the decay of free neutrons has shifted the p/n ratio to 87/13. Out of every 200 particles, 26 neutrons will combine with 26 protons to form 13 helium nuclei. This leaves 148 protons, so the mass ratio of the nuclei produced is:

$$\frac{13 \times 4}{200} = 26\%$$

in very good agreement with the observed fraction of helium in the universe. This is one of the most important experimental confirmations of the big bang theory.

Nucleosynthesis stops when the temperature has dropped below that required for nuclear reactions, which happens 13 minutes into history.

12.3.6 Recombination

Nothing much happens now for the next 300000 years or so. The universe continues to expand and cool until, coincidentally, two events happen at more or less the same time.

The universe has expanded to such an extent by now that the density of radiation drops below that of matter and the further expansion becomes governed by matter. The expansion rate shifts from being $S \sim t^{1/2}$ to $S \sim t^{2/3}$, t being the time after the big bang.

Up to this time the energy of photons in the universe has been sufficient to ionize any hydrogen atoms that form from protons and electrons. Now the photon energy drops below this value and the electrons rapidly combine with protons. When this happens the universe stops containing large numbers of free electrical charges. There is nothing left for the photons to interact with and they decouple from matter. The universe is said to have become *transparent*. This event is called the *recombination* of electrons (although strictly speaking if we are moving forwards from the big bang, these electrons have never combined before!). The photons, meanwhile, do not disappear. They continue to exist in the universe and are steadily red shifted by the expansion. Some 15 billion years later, they are detected by a microwave antenna and then become known as the microwave background.

12.4   The future

The future of the universe is determined by the amount of matter in it.
Friedmann's equations predict three possible outcomes (see figure 12.6).
If the density of the universe is greater than a critical value, then
eventually the expansion will stop and then reverse itself. The universe
will start to contract back into some big crunch in the future.

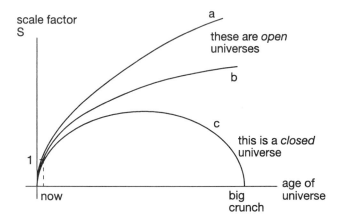

Figure 12.6 Possible futures of the universe. a, density $<$ critical value;
b, density $=$ critical value; c, density $>$ critical value.

If the density is less than the critical value, then the expansion will
continue for ever at a rate that slowly decreases towards some fixed
value. If the density of matter is equal to the critical density, then the
universe will also continue to expand for ever, but the rate of expansion
will become indistinguishable from zero; this is the Einstein–de Sitter
model mentioned before.

Clearly it is of critical importance to measure the density of the universe.
This is not an easy task. Current measurements of the visible matter
suggest that the density is a lot less than the critical value. However,
there is probably a great deal of matter that cannot be seen (as it does not
shine). Neutrinos, for example are certainly present in huge numbers.
If it is shown that neutrinos have mass, then their total contribution to
the universe's mass may be enough to bring it up to the critical value.
We will take up this point in much more detail in the next chapter.

12.5 Summary of chapter 12

The big bang idea rests on three pieces of observational evidence:

- Hubble discovered that the light from distance galaxies was shifted towards the red end of the spectrum and that the size of the shift was proportional to the distance to the galaxy;
- the observed amount of helium in the universe is too great for it to have been manufactured inside stars;
- there is a uniform microwave background radiation characterized by a black body spectrum of approximately 3 K.

- The universe is in a state of expansion in which the space between galaxies is being stretched;
- light waves crossing this space will also be stretched, explaining Hubble's law;
- if the universe is now expanding it follows that at one time it was much smaller than it is now;
- at earlier times the universe was so dense and hot that particle interactions took place commonly;
- the temperature of the universe can be characterized by the energies of the particles in it;
- various processes keep the particles in thermal equilibrium above critical temperatures;
- below its critical temperature a particle stops being a major constituent of the universe;
- above the critical temperature the particle and its antiparticle are kept in thermal equilibrium;
- GUTs suggest that as the critical temperature for X and Y particles is crossed there may be a small excess of matter over antimatter formed;
- helium is manufactured in the early stages of the universe in just the right amount to agree with the current measurements;
- after hydrogen atoms form the photons of the universe decouple and cool due to the expansion to become the 3 K background;
- the fate of the universe depends on the density of matter compared with a critical value that can be calculated;
- if the density is $\leq$ the critical value then the universe will expand forever (an open universe);

- if the density is > the critical value then the universe will eventually recollapse (a closed universe).

Notes

[1] You are probably aware that each atom has its own pattern of wavelengths that it emits, e.g. sodium strongly emits a pair of wavelengths in the yellow part of the spectrum. However, a solid block of sodium when warmed will glow red like any other material. This is because the atoms of sodium are bonded together, which has the effect of turning the line spectrum of the individual atoms into a continuous emission over a range of wavelengths. The same sort of thing is happening inside stars.

[2] Joseph von Fraunhofer (1787–1826).

[3] It is a little known fact that the heavier elements of which we are made were first built out of hydrogen and helium inside a star that exploded billions of years ago scattering its debris across the galaxy. It is from this debris that our planet and our bodies have been formed.

[4] Now that the aptly named orbital Hubble telescope has been repaired it can resolve individual stars in distant galaxies which will help astronomers to measure the distances to galaxies more accurately. Distance measurements are a great problem in astronomy and one of the major reasons why H is so difficult to determine.

[5] In order to avoid confusion between kinetic energy and temperature I have used KE in these equations instead of the more normal T.

[6] This corresponds to the 'flat' model of the universe. The theory of inflation (see chapter 13) would imply that the universe is either exactly flat or very nearly so.

[7] I did say that it was a *small* excess.

Chapter 13

Latest ideas in cosmology

The standard big bang model as outlined in the previous chapter has been highly successful in explaining many of the features of the universe that we see. However, as we have become more accustomed to working with the big bang and as new observations have come in from the Hubble space telescope and other modern equipment, the theory has been tested ever more strenuously. Hardly surprisingly there are some areas that are in need of refinement. In this chapter we shall take a look at some of the problems that are of concern to modern cosmologists and the ideas that they are working on to try and solve them.

13.1　How do galaxies form?

The Friedmann models assume that matter is distributed smoothly throughout the whole universe. This is an excellent approximation in the early moments of history, and appears to hold when we examine the distribution of clusters of galaxies at the scale of the whole universe. However, when we look at the universe at smaller scales we find that it is rather lumpy; there are planets, stars, galaxies and clusters of galaxies. The universe is not uniform at every scale. It has immense structures within it.

When astronomers turned their attention to the details of galaxy distribution in the universe they found that they congregated into clusters. At a higher level of structure clusters merge together into superclusters. These are huge regions of the universe in which the number of galaxies per unit volume is far higher than the average. Often these superclusters are strung out in long filaments.

Between these superclusters are gigantic volumes known as voids that contain virtually no galaxies. Typically these voids can be 150 million light years across and contain ten times less than the average number of galaxies. Figure 13.1 shows the results of a survey of galaxy distribution in a slice of the sky. The large voids and filament-like regions where galaxies congregate are clearly visible.

Pictures like this have been compared to a collection of soap bubbles sticking together. The galaxies are grouped along the surfaces where the bubbles meet and the large voids are inside the bubbles. To carry the bubble analogy further, we can imagine a raft of bubbles on the surface of a bath of water. Viewed by a person standing outside the bath, the bubbles seem to be a uniform white foam, but up close there is a great deal of non-uniformity and structure. The same is true in the universe.

As yet we cannot explain fully why these structures have formed. We are not even sure whether matter clumped together on a small scale to

Figure 13.1 (*See opposite page.*) This 3D computer generated image illustrates the existence of clumpy filamentary structure in the distribution of galaxies (spheres) on large scales in the universe. This 'slice' of the universe covers an area in the sky about 100 degrees in length times 35 degrees in thickness (corresponding to roughly 700 million light years by 250 million light years). We are positioned at the leftmost corner of the pie-shaped slice, and the distances of the galaxies have been determined by measuring their red shift (or velocity) with a spectrograph on a large telescope. The elongated radial feature in the centre is the Coma cluster of galaxies. The longer feature across is known as the Great Wall, and is one of the largest known structures in the universe. (Courtesy of Lars Lindberg Christensen, Bo Milvang-Jensen and Michael J D Linden-Voernle, Niels Bohr Institute for Astronomy, Physics and Geophysics, Astronomical Observatory, Denmark.)

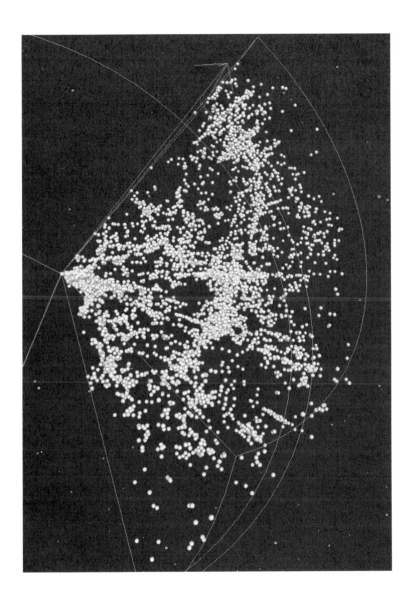

form galaxies that were then drawn together into clusters, or whether the large volumes formed first and galaxies condensed within them. It is a great mystery.

One thing is clear, however. The only way for these structures to form is if the matter in the early universe was not quite as smooth as assumed in the Friedmann models.

In the very early universe any regions of space that had a slightly higher density than average would expand slightly more slowly (see figure 13.2) than the rest (held back by the gravity within them). The density of matter in the space surrounding such a region would decrease as the universe expanded. As long as the density of matter surrounding the region is more than twice the density within it, the gravity of the surrounding matter is pulling outwards on the matter inside sufficiently to stop it collapsing under its own gravity. Eventually the outside drops to less than half the inside density and at this point the matter inside the region starts to collapse inwards. As the matter collapses its density

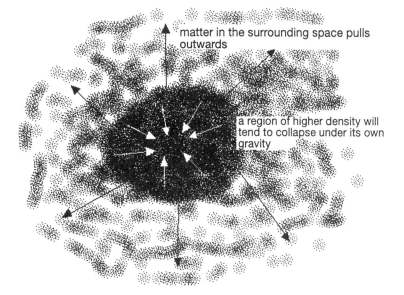

matter in the surrounding space pulls outwards

a region of higher density will tend to collapse under its own gravity

Figure 13.2 A region of slightly higher than normal density will expand more slowly than the surrounding space.

increases. This causes the gravitational effect to grow. The region is in a state of catastrophic collapse and, if small enough, will form galaxies and the stars within them.

A great deal of recent effort has been spent trying to find evidence for these small density fluctuations in the early universe. Evidence for them should be imprinted on the cosmic microwave background radiation.

Until the epoch of recombination (about 300000 years after the big bang) electromagnetic radiation was kept in thermal equilibrium with matter as there were plenty of free charges about to interact with the radiation. After recombination there were no more free charges and the photons decoupled from the matter. Since then they have been steadily red shifted and now form the cosmic microwave background.

If we measure the temperature of the background radiation and produce a map of the sky showing temperature variations this will also be a map of the density variations that were present at recombination. Results obtained from the Cosmic Background Explorer (COBE) satellite show that there are fluctuations in the temperature of the background radiation between different parts of the sky (see figure 13.3). The satellite is sensitive enough to inspect patches of the sky about 0.02° across and compare patches in temperature. The fluctuations measured are very small—*no bigger than a hundred thousandth of a degree*. Furthermore, the regions in which the temperature seems constant are huge—about 100 million light years across.

This presents two problems. How can galaxies and clusters of galaxies form from such a smooth background, and how come the distribution of matter in the universe happens to be so smooth? The smoothness problem is taken up in the next section.

Cosmologists can take the COBE sky map and use it to estimate the size of the regions in the universe that had higher than average density at the time of recombination (the last moment at which the structure of matter could imprint itself on the radiation). As the universe expands these regions grow in size. Detailed calculations show that the ratio of the density in these regions to that in the outside space increases by the same factor as that by which the universe's temperature decreases over the same period.

COBE–DMR Map of CMB Anisotropy

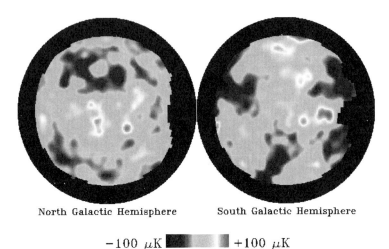

North Galactic Hemisphere   South Galactic Hemisphere

$-100 \ \mu K$ ▮▮▮▮ ▮ $+100 \ \mu K$

Figure 13.3 The COBE sky map. The COBE datasets were developed by the NASA Goddard Space Flight Center under the guidance of the COBE Science Working Group and were provided by the National Space Science Data Center (NSSDC) through the World Data Center A for Rockets and Satellites (WDC-A/R&S).

Minimum density variation at recombination

We can write the ratio of the inside density to the outside density as:

$$\frac{\text{density inside region}}{\text{density in surrounding space}} = 1 + x.$$

Now, as the universe expands the value of x increases in proportion to the change in temperature.

temperature at recombination = 3000 K

temperature now = 3 K

therefore x is now 1000 times bigger than it was at recombination.

The higher density region will start to collapse if the ratio of the inside density to the surrounding space = 2, therefore in order for collapse to start x must equal 1.

> (*Continued.*)
>
> If regions are starting to form galaxies now, $x = 1$ in these regions implying that at recombination $x = 1/1000$, therefore the variations in density that we should see by looking at the COBE results should be at least $1/1000$ or galaxies could not have formed by now.

Unfortunately this is far greater than any variation seen by COBE.

The situation is actually much worse than this. We have just calculated the *minimum* density variation required to start galaxy formation now. Yet when we look into space we see quite a lot of galaxies already formed! These must have been triggered by much larger density fluctuations at recombination which grew to the collapse point up to 10 billion years ago. This places them way outside the limits set by COBE. The problem can be eased by assuming the existence of *dark matter*.

13.2 The evidence for dark matter

Most of the matter that we see in the universe is in the form of light-producing objects—stars. Some matter we can see because it gets in the way of the light from stars (dust clouds and nebulae). Some we can see because it reflects star light (the planets in our own solar system for example). It is quite possible however, that there are large amounts of matter out there that do not interact with electromagnetic radiation. This would mean that we could not detect it in any normal way. Experimental evidence suggesting that there is some form of *dark matter* that we have not yet identified in the universe is growing. The presence of dark matter can be inferred in three ways: from the gravitational effect that it has on the motions of stars in galaxies, the motion of galaxies within clusters and the expansion of the universe.

13.2.1 The motion of stars in galaxies

The stars in a spiral galaxy (see figure 13.4) are rotating about the centre of the galaxy. They are held in orbits by the gravity of other stars that are nearer to the centre of the galaxy.

Figure 13.4 A typical spiral galaxy. (Image from the Ohio State University Bright Galaxy Atlas (Frogel *et al*), © 1996, The Ohio State University Research Foundation, reprinted with permission.)

Astronomers can study the rate of this rotation by looking at the Doppler shift of the light coming from various parts of the galaxy[1]. As the stars move towards us their light is Doppler shifted towards the blue part of the spectrum. As they move away, the light is red shifted. Comparing light from one edge of the galaxy with the other, and also the centre of the galaxy, can establish the rotation curve for the stars within it. Figure 13.5 shows a typical rotation curve, a graph showing the speed at which stars are travelling against distance from the centre of the galaxy.

Simple gravitational calculations show that the speed of stars should

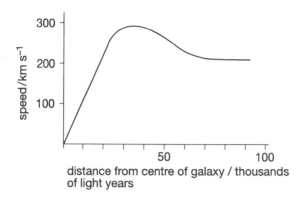

Figure 13.5 The rotation curve for a typical spiral galaxy.

decrease as you move further from the centre. The speed at which a star can remain in orbit is determined by its distance from the centre of the galaxy and the mass of stars pulling it in towards the centre. As you move out through the central bulge of the galaxy the speed increases, as there are more and more stars pulling an individual in. However, once you move out of the bulge into the outlying regions, the density of stars decreases and the number pulling a star inwards stays roughly constant as you get further out. Now one would expect the speed to decrease with distance as the pull of gravity from the stars in the central bulge decreases the further one gets from them.

But it does not continue to drop off; it stabilizes to a virtually constant value. This would be an indication that there was more mass present at the edges of the galaxy. However, a simple visual inspection shows that this is not true—unless the mass is invisible.

The stars are also moving too quickly. Computer simulations of galaxies with stars moving at typical measured speeds show that the galaxy is unstable and quickly starts to break up. There is not enough mass to hold it together at such high speeds.

Both effects can be explained if the galaxy is surrounded by an enormous halo of matter that extends for many times the visible radius of the galaxy (see figure 13.6). As we do not see such a halo in our telescopes it must be of very dark matter (i.e. it does not interact with or produce much light of its own). Estimates of how much dark matter is required in this halo indicate that there needs to be between two and four times the amount of visible matter[2]. The part of the galaxy that we can see represents only a quarter of its total mass.

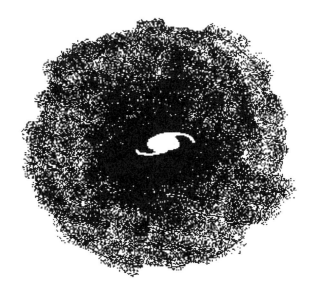

Figure 13.6 A halo of dark matter surrounds a galaxy holding it together by gravity.

13.2.2   The motion of galaxies in clusters

Astronomers can also look at the red shift of light from galaxies within a cluster and measure the speed of the whole galaxy relative to its partners. Our own galaxy is a member of a small cluster that includes the nearest galaxy to ours, Andromeda (22 million light years away), the large and small Magellanic clouds (galaxies visible in the southern hemisphere and first observed by the explorer Magellan) as well as something like twenty other small galaxies. Measurements indicate that Andromeda is moving towards our galaxy with a speed of 100 kilometres per second.

This is an uncomfortably large speed. At this sort of speed Andromeda would have long ago left the local cluster if not for the gravitational pull of the other members. Most of this pull is coming from the Milky Way. If we compare Andromeda's speed to the stars in our galaxy, then problems start to arise. Our sun, for example, orbits the centre of the Milky Way at a distance of 20000 light years and with a speed of about 200 kilometres per second. Andromeda is 1100 times further from the centre of our galaxy than the sun is, and moving at half the speed, yet there is still enough gravity at this distance to hold it in the group. Something is very wrong here. Clearly there is a great deal of mass in our galaxy that is further away from the centre than we are[3].

If we were to treat the Andromeda galaxy as a very large star and plot its motion on the rotation curve of our galaxy, then to explain its speed we would have to assume that the halo of dark matter surrounding our galaxy extended millions of light years beyond the edge of the visible galaxy (or about 70 times the visible radius) and *contained ten times the visible amount of matter.*

The story does not stop here. Our local group of galaxies is apparently being pulled towards a large supercluster in the constellation of Virgo[4]. This is an enormous grouping of thousands of galaxies about 60 million light years away. Measurements suggest that the Milky Way is moving towards the Virgo supercluster at a speed of 600 kilometres per second. This indicates that the total mass of the supercluster could be up to 30 times the visible matter.

It seems that every time we increase the scale of our observations we see evidence for dark matter. Not only that but the proportion of dark matter increases with the scale. This is not too upsetting as it is only the amount of matter between objects that affects their motion. However, it does suggest that each galaxy may only be a tiny pin point within a truly enormous volume of dark matter—possibly stretching tens of millions of light years (over a thousand times the visible radius).

It has taken astronomers a little time to get used to the idea that telescopes may only be showing them a small fraction of the matter present in the universe. The evidence for dark matter from galactic motion seems well established. Even so there might be room for some doubt but for the evidence obtained from the expansion of the universe.

13.2.3　The expansion rate of the universe

The Hubble constant should not be constant. If we look at the farthest galaxies that we can see, then we are also looking at very old galaxies (as the light has taken a long time to cross the distance). At earlier times in the universe's history the galaxies should be moving more quickly as gravity has not had as much time to slow them down. The Hubble constant calculated for a set of nearby galaxies should have a smaller value than that calculated for a set of distant galaxies.

It is notoriously difficult to measure the Hubble constant, so measuring its change is even harder. At the moment the best measurements that we can provide place the deceleration parameter (the ratio of the Hubble constants) somewhere between 0.2 and 2. From this, admittedly crude, measurement, cosmologists can estimate the amount of matter in the universe that is providing the gravity to slow the galaxies down. The estimate turns out to be far greater than the amount of visible matter.

13.3　Back to galaxy formation

Having now surveyed the evidence for the existence of dark matter we can return to the original question of galaxy formation.

The ripples in the cosmic microwave background indicate that the distribution of matter in the universe at recombination was too smooth to trigger the formation of galaxies. Dark matter can provide the basis for a solution to this problem.

The dark matter may have started to form large clumps of higher than average density some time before recombination. If the dark matter is of a form that does not readily interact with light (as seems likely—our galaxy is presumably inside a giant halo of the stuff, and we can see out!), then there would be no evidence of this imprinted on the microwave background. These clumps of dark matter could act as traps that localize ordinary matter and trigger galaxy formation. To proceed any further with a theory one has to identify what the dark matter is, as the way in which it clumps together depends on its nature.

Two distinct classes of dark matter have been considered: *hot dark matter* and *cold dark matter*. Of course the terms do not refer to temperature in the conventional sense. Hot dark matter is assumed to be moving at close to the speed of light when the clumping process starts. Cold dark matter, on the other hand, was moving at speeds very much slower than light as it started to form clumps. There is no conclusive evidence that can directly show what the dark matter in the universe is, so the various candidates have to be examined theoretically with the aid of computer models. The purpose of these pieces of software is to simulate part of the early universe. Various mixes of dark and ordinary matter are programmed in and the simulated universe is allowed to expand. As the programme proceeds galaxies start to form from the clumped dark matter. Astronomers are looking for the mix that gives the most realistic simulation of the of the superclusters and voids that we see in the real universe.

In the early 1980s hot dark matter was all the rage. Neutrinos were studied in the role of hot dark matter. Calculations had shown that there should be several hundred neutrinos per cubic centimetre left over from the big bang. Even if neutrinos had a very small mass, in total they would contribute an enormous mass to the universe. Furthermore they only interact via the weak force. Being neutral they cannot interact with electromagnetic radiation. They satisfy all the requirements for dark matter.

Just at the time that cosmologists were in need of a form of dark matter, the particle physicists found massive neutrinos that would do the job. The results of experiments to measure the mass of the electron-neutrino were suggesting that they had a mass in the region of 30 eV/c^2. These experiments subsequently proved impossible to reproduce and the question of neutrino mass is still open. At the time, however, they were another piece of persuasive evidence. An industry grew up overnight as simulated universes populated with massive neutrinos were put through their paces.

Unfortunately, they did not work out. With such a small mass in each neutrino they would still be moving at essentially the speed of light. This means that only very large clumps of neutrinos have enough mass to hold together gravitationally. In a universe dominated by hot dark neutrinos the density variations would be regions of mass equivalent to a hundred trillion stars (about a thousand galaxies). Any smaller regions would rapidly be smoothed out by the fast moving neutrinos[5]. With regions of such size collapsing under their own gravity the largest structures form first—the superclusters. The problem is that it takes too long for the neutrinos to clump together on smaller scales to form galaxies within the clusters (the initial cluster would be a huge cloud of gas). In a hot dark matter universe, galaxies should only be forming about now. So, after a brief period of success and hope, interest in hot dark matter started to fade as physicists realized that they could not explain galaxies in this way.

Inevitably interest turned to cold dark matter. Unfortunately, there is no single particle that is an obvious candidate for this role. Particle physics theorists have come up with a variety of exotic objects that have the general features required. The collective name coined for the various candidates is WIMPS (weakly interacting massive particles). The requirements are that the particle should not be able to interact electromagnetically, it has to be very massive (and hence slow moving) and be generated in sufficient quantities in the big bang. There is quite a list of candidates.

Axions

Axion particles were suggested in the late 1970s as a way of tidying up some aspects of QCD. If they exist, axions should have an incredibly small mass, of the order of 10^{-5} eV, but they would have been produced in copious quantities in the big bang. Furthermore, the processes leading to their production would ensure that they were nearly stationary (even though they are very light in mass).

Photino

The supersymmetric partner of the photon. None of the supersymmetric particles have their masses fixed by the theory, but the fact that they have not been observed yet suggests that the lightest of them (thought to be the photino) must have a mass of at least 15 GeV.

Quark nuggets

The ordinary hadrons that make up the visible matter in the universe are composed of u and d quarks. The s quark (and the others) are too massive and quickly decay. However, it has been suggested that a nugget of matter produced in the big bang with roughly equal numbers of u, d and s quarks might just be stable. The key here is the size. Quark nuggets would be composed simply of quarks packed close together and could be between 0.01 cm and 10 cm in size. This would give them a mass between 10^3 and 10^{15} kg!

Baryonic matter

In the terminology of cosmology baryonic matter is any ordinary form of matter made from protons and neutrons. The visible matter that we can see in the universe and the dust clouds and nebulae that we can infer are all baryonic matter. It is possible that there is more of this stuff in the universe than we realized and that it contributes to the mass of dark matter. Detailed calculations show that the correct proportions of hydrogen, helium, lithium and deuterium produced in the nucleosynthesis era of the big bang can be obtained with a total mass of baryonic matter in the universe up to 10 times that which we observe. This very neatly fits with the estimate of dark matter obtained from rotation curves. Perhaps all the dark matter needed to explain rotation curves is ordinary baryonic stuff that for some reason did not light up. There are several possibilities.

Perhaps galaxies are surrounded by clouds of hydrogen that are not big enough to block out the view, but there are enough of them to have a gravitational influence.

Recent observations suggest that there may be more brown dwarf stars than we had previously thought. Brown dwarfs are star-like objects that were not large enough to heat up sufficiently to light nuclear reactions. They are failed stars. There may be enough of them to explain most of the baryonic dark matter (although it is not clear why they should be distributed in a halo round the galaxy when all the successful stars are contained within it).

Black holes may also be contributing.

Black holes (as virtually everyone must know by now) are the remains of very massive stars that have collapsed under their own gravity. The gravitational forces of these objects are so great that not even light can escape from them. They may be enormously massive, yet smaller than the earth.

Steven Hawking suggested that mini black holes (smaller than atoms) could have been created in the big bang. It is possible that they could be contributing to the dark matter. However, Hawking also showed that such mini black holes should eventually explode in a gigantic burst of γ rays. No evidence for this has been seen, so it is doubtful that they exist.

The consensus among astronomers and cosmologists is that there may be enough baryonic material about to explain the rotation curves, but something far more exotic is required to provide the rest of the dark matter that is apparently there.

This is not the total list of candidate cold dark matter objects, but it does show the sort of ingenuity that is being applied to the problem.

Cold dark matter started to come into vogue in 1982. Like hot dark matter the early results were promising. The particles were much more slowly moving and so could clump at far smaller scales—about the size of galaxies. However, cold dark matter also ran into a problem. The reverse problem in fact. Although cold dark matter universes had no trouble producing galaxies, they could not generate the huge superclusters and voids. For a while some physicists hoped that the voids that had been seen were just flukes and not common features of the universe. However, as more and more measurements and charts came in from the observers it became increasingly clear that filaments and voids were a common feature of the universe. Cold dark matter does not work either.

And that is more or less where we are now. The ingenuity of the theorists has not rested, however. It is possible to tinker with both the cold and hot dark matter models. However, to continue with this story we need to introduce a second line of development.

13.4 Problems with the big bang theory

13.4.1 The horizon problem

Measurements of the cosmic microwave background radiation show that the equivalent temperature is virtually the same at every point in the sky to a very high degree of accuracy. As we have seen, the smallness of these variations poses problems for people trying to understand how galaxies can form. However, the smallness of the variations is a problem in itself. The universe is far too smooth.

As the universe expanded points on opposite sides of the sky have always been a long way from each other. When the universe was very young, they may only have been a few centimetres apart. However, at such an early stage of the universe's history the expansion rate was very rapid. A ray of light heading across the universe has had further and further to travel as expansion takes place. It might have taken light 15 billion years or more to cross the ever expanding distance, even though the points started off when the light was emitted only a few centimetres apart.

The diagram in figure 13.7 shows two objects that are an equal distance, d, from us. If the objects are not too far apart in the sky ($\theta < 1$ rad) then:

$$x_0 = d\theta_0$$

the subscript 0 indicating a quantity measured in the current epoch.

Let us assume that these objects are so far away that light emitted from them at the moment of recombination is only just reaching us now (the objects can be taken to be different spots on the COBE sky map).

If this is so, then the time it has taken for the light to reach us is the age of the universe now, t_0, minus the age of the universe at recombination, t_r:

$$t = (t_0 - t_r)$$
$$\sim t_0$$
$$\text{as } t_0 \gg t_r$$

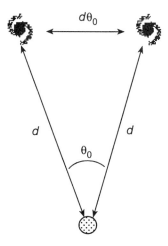

Figure 13.7 The angular distance between two objects is related to the linear distance.

(compare 15 billion years with 300000 years!). The light has been travelling to us for this time at the speed of light, c, so:

$$d = ct_0$$
$$\therefore \quad x_0 = \theta_0 t_0 c.$$

In the previous chapter we saw that it is convenient to scale distances using the parameter S. It is generally assumed that $S(t_0) = 1$ and so $S(t < t_0) < 1$. Using this we can calculate how far apart the two objects were when recombination took place. If x_r is the maximum distance between objects at the time of recombination:

$$x_r = Sx_0$$
$$= S\theta_0 t_0 c.$$

Until the time of recombination, the particles in the two objects were able to interact with electromagnetic radiation. (After this time the free charges formed atoms and the universe became transparent.) If the objects exchanged photons, then some form of 'communication' was possible between them. This, of course, is provided that they were close enough for light to have time to cross from one to the other. So, we can calculate the maximum distance between objects that could communicate as the maximum time available is the time between the big bang and the moment of recombination. Thus

$$x_r < ct_r$$
$$\therefore \quad S\theta_0 t_0 c < ct_r$$
$$\therefore \quad \theta_0 < \frac{t_r}{St_0}.$$

From the previous chapter, the temperature of the universe also scales with the parameter S:

$$T \propto 1/S.$$

Using this we can estimate θ_0:

$$\theta_0 < \frac{t_r T_r}{t_0 T_0}$$
$$= \frac{10^5 \text{ yr} \times 3000 \text{ K}}{10^{10} \text{ yr} \times 3 \text{ K}}$$
$$= 10^{-2} \text{ rad}$$
$$\sim 1°.$$

This calculation shows that any two objects in the sky that are further than 1° apart cannot have interacted with each other before recombination (after all, nothing can travel faster than light and light would not have had the time to make the journey). If this is the case, *then there is no physical mechanism that can ensure that these two objects are at the same temperature.*

Yet the COBE results found objects much further apart than this at virtually identical temperatures. This is known as the horizon problem—how can two places in the universe be the same temperature (in equilibrium) if one is 'over the horizon' from the other?

13.4.2 The flatness problem

In the previous chapter we discussed how the density of the universe will determine its future. It is easy to see that the density of the universe changes with time as there is a constant amount of matter in an increasing volume of space. It is also quite reasonable to suppose that the density of matter that would be required to close the universe (the critical density) also changes with time. However, it is not quite so obvious that these two quantities should evolve differently as the universe ages.

The density of the universe varies by S^{-3} as the volume of space increases as S^3. The variation of S with time depends on the type of universe being considered. In an open universe S increases like $t^{2/3}$ when t is quite large. In a closed universe S will increase to a maximum value and then decrease again as the big crunch approaches.

The critical density is the amount of matter per unit volume required to provide enough gravitational pull to halt the expansion of the universe. As the force of gravity depends on distance, the amount of matter required changes as the universe expands—not only does the density of a constant amount of matter decrease, but its gravitational pull on parts of the universe further away also decreases as space gets bigger. The upshot of this is that the amount of matter now required to close the universe is different to what it would have been several million years ago, or indeed what it will be at some time in the future.

So, the density of the universe and the critical density vary with the age of the universe. The ratio density/critical density, Ω, is not constant. If density and critical density evolved in the same way, Ω would not change. However, in both open and closed universes it increases with time. This does not mean that we could have switched from being in a closed to an open universe at some point in the past. As Ω evolves it will always stay > 1 if it started that way, and it will always be < 1 if that was the initial value. The fate of the universe was determined at the moment of the big bang. The interesting question is, what was the ratio initially?

It is very difficult to measure the amount of matter in the universe. Astronomers have made several attempts and the largest value of Ω that has been estimated is 0.2 (i.e. the ratio *now* is 0.2, it would have been different in the past). This indicates that we are living in an open universe. Now 20% of the matter required to close the universe does not seem to be that close to the critical value. However, it is in fact remarkably close and is a cause for some concern among cosmologists.

The problem is that a value of Ω that is 0.2 now implies that Ω must have been very close to 1 at the moment of the big bang. Ω gets bigger with time very rapidly. To be that close now, after this rapid increase, implies that it must have been fantastically close at the big bang.

Figure 13.8 shows how Ω increases for open ($\Omega < 1$), closed ($\Omega > 1$) and flat ($\Omega = 1$) universes.

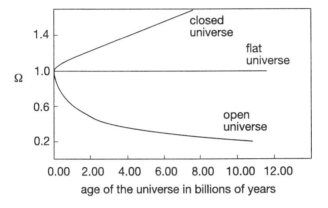

Figure 13.8 The evolution of Ω.

If Ω is exactly 1 at the initial moment, then it stays exactly 1 for ever and the universe expands in a regular manner (the flat universe). If, however, it is just a bit less than 1 or just a bit more, then it rapidly increases as shown. The curves are plotted for a Hubble constant of 23 km s^{-1} per million light years.

If we project a value $\Omega = 0.2$ back to 10^{-49} seconds after the big bang, then at that moment Ω could not have been different from 1 by any value greater than 10^{-50}. This is extraordinary precision in the initial density of the universe. Seeing this cosmologists suspect one of two things. Either the universe was launched with Ω *exactly* 1 or there is some physical law at work that drives Ω towards 1 at some point after the universe starts.

This is known as the flatness problem. Why should the universe have been launched with a density so incredibly close to the single unique value that would make it flat[7] for all time?

13.5   Inflation

Late on 6 December 1979, Alan Guth (then a post-PhD researcher at SLAC) turned his hand to applying GUT theories to the early universe. The next morning he cycled back into the lab clutching his notebook. At the top of one of its pages he had scribbled 'spectacular realization'. His calculations, concluded late the previous evening, had shown him a way of solving the horizon and flatness problems in a manner that arose very naturally from particle physics. The grand co-operation between the very large and the very small had paid off in a dramatic way. Since then Guth's ideas have been refined but the central realization remains.

At the time when the universe cooled to the temperature at which the GUT force broke up into the strong and electroweak forces, the universe must have undergone an extraordinary period of expansion—far faster than anything that had occurred up to that point. So fast, that the universe would double in size every 10^{-34} seconds. Guth termed this rapid expansion *inflation*.

The period of inflation probably started at about 10^{-35} s into the history of the universe and stopped at 10^{-32} s. However, in this minute period of time the size of the universe would have increased by a factor of 10^{50} or more! That would be equivalent to an object the size of a proton swelling to 10^{19} light years across. At a stroke this solves the horizon problem.

Assuming that inflation starts 10^{-35} seconds after the big bang, then in that time light would have travelled a distance of 10^{-35} s $\times$ 3 $\times$ 10^8 m s^{-1} = 10^{-27} m. This would be the maximum distance between points that were at the same temperature at the moment inflation started. Inflation would expand that region of space by a factor of 10^{50}. After inflation the region would be 10^{23} m in size. From then on it would continue to expand in the ordinary Friedmann manner. The key point is the size of that region compared to the visible universe.

We can estimate the size of the visible universe at the moment when inflation stopped by taking the scale factor back to that moment. If the universe is 15 billion years old now, then it is about 15 billion light years across. If the universe is at the critical density after inflation we can use the scale parameters of the Einstein–de Sitter model, as quoted in the last chapter:

While matter dominates the universe

$$S \sim t^{2/3}$$

so

$$\frac{S_0}{S_r} = \left(\frac{t_0}{t_r}\right)^{2/3}$$

where 0 indicates the current time, and r indicates the time of recombination, which is about the same time as the universe switches to being matter dominated. Therefore

$$\frac{S_0}{S_r} = \left(\frac{15 \text{ billion years}}{300000 \text{ years}}\right)^{2/3}$$
$$= 1.36 \times 10^3$$

so the size of our universe at the time of recombination is

$$= \frac{15 \text{ billion light years}}{1.36 \times 10^3}$$
$$= 1.11 \times 10^7 \text{ light years.}$$

Now we need to scale this down further from the time of recombination (300000 years) to the time at which inflation stopped (10^{-32} s). During this time the universe is radiation dominated, so:

$$S \sim t^{1/2}$$
$$\frac{S_r}{S_i} = \left(\frac{t_r}{t_i}\right)^{1/2}$$
$$= \left(\frac{9.46 \times 10^{12} \text{ s}}{10^{-32} \text{ s}}\right)^{1/2}$$
$$= 3.08 \times 10^{22}$$

so the size at the end of the inflation period is

$$\frac{1.11 \times 10^7 \text{ light years}}{3.08 \times 10^{22}} \sim 3 \text{ m.}$$

While this calculation only gives a rough estimate, one thing is clear: our whole visible universe would be a microscopic speck in the region that inflated.

This is why the cosmic microwave background is so smooth—the horizon problem is solved. Because such a small region in the early universe can inflate into such a huge region we would not expect to see any large variations in temperature across a region many times larger than the whole visible universe!

The reason that the cosmic microwave background is so smooth is that the whole of our universe started out as a region small enough to be in equilibrium. Equilibrium was established *before* inflation and was undisturbed by the process.

The same inflationary process also has the effect of driving Ω towards 1. Imagine viewing a balloon as it inflates. The sides of the balloon get flatter and flatter as it increases in size. The effect of this extraordinary inflation is to smooth the universe to flatness.

13.6   When is a vacuum not a vacuum?

The root cause of the inflationary expansion is the Higgs field introduced by theoreticians to explain how the W and Z particles have mass.

Recall that at high energy the Higgs field cannot be seen, but as the interaction energy drops, so the interaction between W's, Z's and the Higgs field gives them an effective mass and the perfect symmetry of the electroweak theory is broken. The weak and electromagnetic forces separate.

A similar mechanism is needed to explain how the grand unified force splits into the electroweak and strong forces. In this case it is the X and Y particles that need to be given mass.

GUT theories incorporate a 'super' Higgs field to do this job. At very high energies the perfect GUT symmetry is in operation. Leptons and quarks cannot be distinguished and all the exchange particles are massless. As the interaction energy drops below the point at which the symmetry starts to break (10^{14} GeV), the interaction between the Higgs field and the X's and Y's give them a mass of the order of 10^{14} GeV. This breaks the symmetry. The X and Y particles become difficult to emit and the quarks and leptons become distinct. As a consequence the strong and electroweak forces separate out.

There is no hope of ever exploring GUTs directly by experiment. X and Y particles lie well beyond the range of any conceivable accelerator. However, early in the universe's history the GUT symmetry must have broken.

In the history of the universe, interaction energy corresponds to temperature and temperature varies with time. The energy at which the GUT split into the separate forces must correspond to a moment in the history of the universe. Guth investigated the effect that this change in the status of the super Higgs field would have on the expansion of the universe.

One of the odd things about the Higgs field is that it can still contain a great deal of energy even if its strength is zero. In all field theories the strength of the field at any point is a measure of how likely it is to find an excitation of the field at that point. If the Higgs field has zero strength, then there are no Higgs particles to be found.

It is a little like the surface of a very deep lake. If there are no ripples on the surface, then the lake effectively 'disappears'. However, when a ripple crosses the surface we can see that the water is there. The Higgs field cannot be seen at high energy as no Higgs particles are being produced. However, as the lake has great depth, so the Higgs field can still contain a great deal of energy even though its presence cannot be detected. This energy is latent within the field and cannot be released as there are no Higgs particles. This energy locked into the Higgs field is the cause of inflation.

The Friedmann equations determine the rate of the universe's expansion by relating it to the amount of energy per unit volume of the universe (the *energy density*).

Before GUT theories were introduced cosmologists only included the energy of the matter and radiation present (radiation being defined as relativistically moving objects, i.e. photons, high energy particles, etc). If there was no energy of this form, then the volume of space was a vacuum in the true sense—nothing present. Solving the Friedmann equations for a matter dominated universe provided the three basic models discussed in the previous chapter. As the universe expands so the energy density decreases (same energy in a bigger volume) and we get a slow expansion with an ultimate fate depending on Ω.

When Guth tackled this problem he realized that the empty vacuum of pre-GUT theory was not empty at all—it contained the super Higgs field. Even through the strength of this field was zero above the GUT temperatures it still contained a great deal of energy and this would have to be included in the energy density of the universe driving the Friedmann equations.

Early in history the energy density of the Higgs field is less than the energy density of the matter and radiation in the universe. The expansion rate is consequently determined by the matter and radiation. However, as the universe expands the energy density of the matter and radiation drops. As the universe reaches the temperature at which the GUT breaks down the Higgs energy density becomes larger than that of the matter and radiation. It takes over controlling the expansion.

This is the moment at which the tremendous period of inflation begins. The Higgs field can drive this rapid expansion as it behaves in a totally different manner to matter or radiation. The energy in the Higgs field does not decrease as the universe expands—it gets bigger!

This marvellous conjuring trick can only really be understood in the theory of general relativity. It relies on the fact that gravitational energy is negative. Empty space containing no matter or radiation is not really 'empty'—if GUTs are to work then they must contain the super Higgs field. As the universe expands the volume of space increases bringing more Higgs field with it. To create Higgs field requires energy—positive energy. As this takes place the energy in the gravitational field gets more negative. The total energy remains the same[6]. The one increases at the expense of the other.

Once this period of inflation has started the difficult part is stopping it. Once again a remarkable property of the Higgs field comes to the universe's aid. At temperatures well above the point at which GUTs break down, the energy in the Higgs field varies with its strength in a manner shown in figure 13.9(a). In order to have minimum energy the Higgs field must be zero everywhere (but note that this minimum energy is not zero).

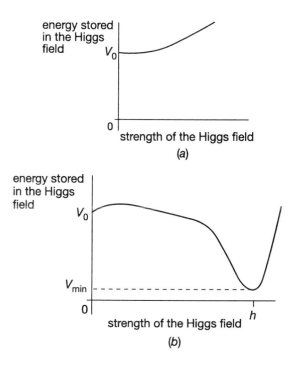

Figure 13.9 (a) The Higgs energy curve at a temperature above the GUT temperature. (b) The Higgs energy curve at a temperature below the GUT temperature.

As the temperature falls so the energy curve of the Higgs field changes to something more like figure 13.9(b). Now the minimum energy is to have a non-zero Higgs field—in other words to create Higgs particles.

Under the influence of this new energy curve the Higgs field starts to change into one with a non-zero value. However, as the curve is very gently sloping near to the zero point the change takes place very slowly—far slower than the rate at which the universe is inflating. As the inflation takes place the new space that is being created is forced to contain Higgs field with an approximately constant value. However, eventually the Higgs field gets to a value quite close to the dramatic down curve in figure 13.9(b) and things start to happen quickly.

The Higgs field suddenly changes to a value h shown on the graph and a tremendous amount of energy is released. All the energy that has been put into the Higgs field (at the expense of the gravitational field remember) is now able to escape and form Higgs particles. The symmetry of the GUTs is broken. Now the vacuum is truly a vacuum as the Higgs energy has gone. The inflationary period stops. The universe has grown by a factor of 10^{50} and it is saturated with Higgs particles of enormous mass. These rapidly decay into more ordinary forms of matter and radiation flooding the inflated universe. The temperature of the universe is increased by this flood of new particles, but not high enough to restore the GUT symmetry. However, the energy is such that shortly after the inflation has stopped the GUT processes result in an imbalance between matter and antimatter being produced (as outlined in the last chapter). When annihilation takes place there will be a net amount of matter left over.

This matter then goes on to form us.

The particles that were present in the tiny pre-inflated volume that went on to form our universe have been spread across a huge volume and their density reduced virtually to zero. They have been replaced by the decay products of the Higgs particles.

This point is very important. The vast majority of the matter that we observe has been created by the decay of the Higgs particles and did not exist before inflation started. The matter that was there initially is a small component of what we see. The significance of this is that the extra matter that inflation has added has driven the density of the universe towards the critical density with very high precision. If the universe inflated by a factor of 10^{50}, then we expect the density of the universe to be at most 1 part in 10^{100} away from being the critical density. However, this does not mean that it will be *exactly* the critical density. If the universe was greater than critical before inflation, then it will still be greater than critical after inflation—but very, very much closer. If it started out less than critical then it will still be less than critical after inflation has happened. From now on the universe will continue to expand in the standard Friedmann manner.

13.7 Ripples

One of the reasons why inflation is so popular with cosmologists is that it explains why the universe is so uniform. However, there is the danger that it does the job too well and produces a microwave background that is even smoother than that observed. For a while this was regarded as a serious flaw in the theory until it was realized that the Higgs field would not be totally uniform everywhere during the inflation. These 'ripples' in the Higgs field would be microscopic variations in its intensity from place to place. The process of inflation would expand them to gigantic size and when the Higgs particles produced from the field decayed into matter they would become density variations in the matter of the universe.

The dark matter, of whatever variety, would also be produced from the Higgs field. Consequently it would have density variations corresponding to the ripples in the field. Ten thousand years or so after inflation had stopped, the Friedmann expansion would have increased the size of these variations (as in section 13.1) and the dark matter started to clump gravitationally. The ordinary matter is still prevented from doing this as the electromagnetic radiation is interacting with the matter and 'blows away' any matter clumps that start to form. However, when recombination has taken place the matter is free from this effect. The COBE results reflect the variations in the matter density set in by the Higgs field, but there are unseen dark matter variations that are much bigger. The ordinary matter is now pulled by gravity to these dark matter clumps.

At this point the story changes depending on the nature of the dark matter. Hot dark matter has been moving so quickly that it has smoothed out some of the density variations set into it by the Higgs field. Only the very largest variations remain. Consequently the ordinary matter will collapse into clouds the size of a thousand galaxies.

From this the individual galaxies form. The problem with this version of the story is that the whole process takes too long. Many galaxies that we observe are too old to have been formed in this way.

In a cold dark matter universe the matter is moving slowly so the same smoothing effect does not take place. When the ordinary matter is gathered onto the cold dark matter clumps it tends to collect in galaxy size masses. Both the dark matter and the ordinary matter are slow moving and tend to mix gravitationally.

In a volume of space containing both types of matter, ordinary matter would soon separate from cold dark matter. The ordinary matter would be able to radiate away energy electromagnetically (triggered by collisions between atoms)—it would slow down. Gravity would then pull this matter into the centre of the cloud where it could condense further and start to form stars and galaxies (see figure 13.10).

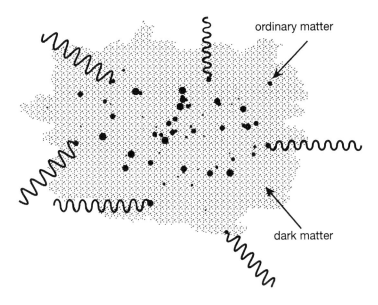

Figure 13.10 Ordinary matter can condense inside dark matter by radiating energy away.

What we end up with here, as confirmed by computer simulations, is a large clump of cold dark matter with a glowing galaxy in the middle.

This is exactly what we see in the evidence for dark matter—the galactic halos. However, there are problems with the cold dark matter idea.

Firstly, cold dark matter is very good at forming galaxies, but it is not very good at forming the very large structures that we see in the galaxy distribution. Secondly, the cold dark matter is very slow moving and so it is difficult to explain why some galaxies are observed with very high speeds.

Finally, estimates of the amount of dark matter seen in galaxies and clusters place Ω at about 0.2 (see section 13.2). However, inflation requires Ω to be very close to 1—far closer than this. So, a great deal more cold dark matter is required. If there are enough clumps of the stuff to bring Ω to 1, then it is difficult to explain why they do not contain galaxies. There are various ways of 'biasing' the production of galaxies so that only the clumps with the very highest density form galaxies, but there is no obvious neat solution to this problem.

13.8 Loose threads

The inflationary model in its modified form outlined above is a very attractive way of solving some of the problems that face cosmologists. To gather the horizon problem and the flatness problem and solve them in one stroke is very persuasive. However, inflation is not without its own problems. The Higgs particle has not been discovered. The type of Higgs energy curve chosen to produce inflation is not fixed by theory, it has been selected to do the job required of it. Furthermore, if inflation is correct then Ω should be almost exactly 1. At the moment the wildest estimates of the amount of dark matter in the universe place Ω at 0.2. There is a great deal of dark matter still to find. This is an extraordinary thing. If inflation is correct, then we do not know what most of the universe is made of. Furthermore, we cannot see it!

One test of the inflationary idea is to measure the temperature variations in the microwave background to a much higher degree of precision than COBE achieved. Inflation relates the variation in matter density of the universe to variations in the strength of the Higgs field from place to place. These can be calculated and definite predictions about the size of the temperature variations in the microwave background can be made. Experiments are being prepared that will improve on the COBE measurements.

Galaxy formation is still a long way from being understood. As observations have improved we have been able to look further out into the universe at fainter and fainter galaxies. The bubbles and voids are still being found. The amount of structure in the universe is incompatible with both cold and hot dark matter ideas. Perhaps some mixture of the two is the answer.

There are other theories of galaxy formation that rely on structures left over from the GUT Higgs field (so-called 'cosmic strings'). As yet nothing has been very successful.

Dark matter is taken for granted now in the cosmological community. It is a standard ingredient in any theory of the big bang. One crucial test would be to find the stuff! This will not be done in observatories. Experiments will either find the supersymmetric particles or not. Delicate experiments are being set up to hunt for the other exotic forms of dark matter. As yet nothing has been found. The theories of the very large are waiting for news from the very small.

13.9   Summary of chapter 13

- The observed distribution of galaxies shows filament-like structures and large voids in which very few galaxies are found;
- the existence of these structures is proving very difficult to explain;
- the COBE map of the microwave background radiation has found very small variations in temperature across the sky;
- the temperature variations reflect differences in the matter density of the universe at the time of recombination;
- these variations are too small to explain galaxy formation;
- the variations are too small as light would not have time to bring all parts of the universe to the same temperature to this degree;
- evidence for dark matter—matter that does not interact with light— can be seen in the motion of stars in galaxies, galaxies in cluster and in the expansion rate of the universe;

- dark matter might help explain galaxy formation as its clumpiness would not show up in the COBE map, and so it could be far more clumped than ordinary matter;
- hot dark matter (neutrinos) does not work as it can only form very large structures;
- cold dark matter (slow moving particles) does not seem to be able to form the large structures that we see;
- inflation solves some of the problems of the big bang;
- inflation is the rapid expansion of the universe that took place in the GUT epoch;
- inflation will force the density of the universe very close to the critical density;
- observations of visible matter place the density of the universe very much less than critical;
- if inflation is right, then most of the universe must be made of dark matter.

Notes

[1] This is a genuine Doppler shift, unlike the Hubble red shift. Astronomers have to subtract out the Hubble red shift in order to find the rotation speeds of the stars within a galaxy.

[2] The estimate of visible matter is made by measuring the light output of the galaxy and comparing it with the light output of a typical star to give the number of stars. Knowing the mass of a typical star gives the mass of the galaxy. The estimation can be refined by including the mass of low intensity stars (so-called brown dwarfs) and the dark dust and nebulae in the galaxy. The answer comes up short of that required to explain the rotation curve by a factor of four. This is much too big a gap to be explained by inaccurate estimation.

[3] If it lies outside our orbit, then it can have no effect on the orbit speed. It is only the mass that lies inside an orbit that counts. That is why the rotation curve was expected to drop away at the edge of a galaxy.

[4] Of course the supercluster is not 'in' the constellation. The stars that make up the pattern we call the constellation of Virgo belong to our own galaxy. These patterns are useful road maps. When an astronomer says an object like a galaxy is in a certain constellation they mean that one has to look out of the galaxy in the direction towards that pattern.

[5] Imagine two points near to each other. One has a large number of neutrinos and the other a smaller number. Random movement between the two points will ensure that the numbers even out over a period of time. Neutrinos are very fast moving objects, so this evening-out process can take place over very large distances. Hence hot dark matter tends to clump together in regions of uniform density that are very large.

[6] It is possible that the total energy of the universe is zero. It has always been a worry that the creation of the universe apparently violates the conservation of energy. However, if the total energy of the universe is zero then the conservation law is obeyed. As the negative gravitational energy increases so the positive Higgs energy increases—still totalling zero. It is the Higgs energy that eventually goes on to form the matter from which we are made. Guth has suggested that the universe might be the ultimate free lunch.

[7] When cosmologists refer to a flat universe they are making a statement about the way general relativity applies to the space within it. For our purposes it is simply a term meaning a universe in which $\Omega = 1$.

Postlude

The future of particle physics

When I started writing this book the LEP experiment at CERN was about to be switched on. Now some years later LEP has turned into LEP2 and an era in particle physics is about to end.

On 2 March 1995 Fermilab announced the discovery of the top quark. Over a two-year period 6 trillion collisions were taken as a basic sample, from which 40 million promising events were sorted. In the final stage 17 events were produced that seemed to involve the top quark. This extraordinary sifting through data shows how difficult particle physics experiments have become. Equally, that we have had to wait 18 years between the discovery of bottom and top demonstrates that sometimes we have to wait for the technology to catch up.

LEP has been an extraordinary success in many ways. That nearly a quarter of the world's particle physicists should work together on designing, building and using an accelerator and its detectors is a triumph of international collaboration on a scale that puts the UN to shame.

LEP was designed primarily to test the standard model. The standard model has survived this in remarkably good order. In addition, persuasive evidence has been gathered that there are only three generations of quarks and leptons.

Yet in some ways LEP has been a disappointment. There are several ways in which the standard model can be extended. Supersymmetry is an idea that has been in currency for many years, yet aside from some tentative hints there is no experimental evidence to support it at present. The Higgs particle has become the most hunted beast in particle physics. Some people hoped to see it with LEP. The upgrade to the machine LEP2 allows it to run at higher energies so perhaps it may turn up there. I, for one, am not taking bets.

The Superconducting Supercollider (SSC) project was to be the next great step forward in particle physics experimentation. However, in 1993 the project was cancelled by the American Congress as its projected costs (approximately $11 billion) were not acceptable in the political climate of the time.

This giant accelerator was to collide particles at 40 TeV energies. The cancellation notice sent a shockwave through the community. It took many people by surprise as $1 billion had already been spent on developing the technology for the accelerator and detectors. For the first time the particle physics community had a very public project cancelled. For a while some people were in such shock they were talking about the end of particle physics. Now the general opinion is that we are going to have to accept that countries will not bank roll ever greater energy accelerators for ever. We are going to have to be cleverer in our experimental design.

A new generation of accelerators is needed to test the GUT theories mentioned in chapter 11. We know that there must be some new physics out there, but we do not have the experimental information to choose which of the options (if any) is correct.

CERN's latest accelerator project is the LHC (Large Hadron Collider). This does not aim to reach the huge energies of the SSC and will run in the same tunnel as the LEP ring does now (saving on construction costs). Even so, the European collaboration has had to fight hard to gain approval for construction.

The intention was that LHC would become operational in 2004 to collide protons at 9.3 TeV. Later, in 2008, a second stage of construction would enable the machine to reach 14 TeV. However, at a meeting in December 1996, the CERN council agreed a drastic change to this programme. Now the accelerator is due to be completed in one stage by 2005. The acceleration in the project has come about due to a substantial series of contributions made by non-member countries (i.e. countries that do not contribute to the annual running budget of CERN) negotiated during 1996:

Japan	8.85 billion Yen;
India	$12.5 million;
Russia	67 million Swiss Francs;
Canada	$30 million in kind;
USA	$530 million in contributions to LHC experiments and the accelerator.

Moving outside the normal CERN circle for obtaining finance has been an important decision, but it was made in the shadow of a demand from the member states for a reduction in the annual contribution. It was decided that the LHC funding should continue as planned, but that there should be a rolling programme of reductions in the total budget for the member states:

7.5% in 1997
8.5% in 1998–2000
9.3% in 2001 and thereafter.

The LHC will become an accelerator of world significance. Although its energy will be still a very long way from the GUT region it is expected that new particles will appear. It would be a major shock if the Higgs particle did not turn up at this energy. As well as this the so-called 'supersymmetric' particles are expected to have masses of the order of 1 TeV. Supersymmetry theory is complementary to GUTs and provides a simple mechanism to introduce gravity into the same theory as the other forces. It is one of the most promising approaches that theoreticians have explored and it would be a major shock if the particles were not to be seen by the LHC.

Moving beyond the GUT energies the latest theories incorporate the supersymmetric ideas and offer tantalizing glimpses of a completely unified theory of physics that incorporates gravity. The most plausible work is being done on superstring theory, but the mathematics involved is extremely advanced and difficult.

From the theoretical point of view, the implications of particle physics for the big bang are decisive in deciding which theories are useful. The COBE results have already ruled out one form of GUT that produces density variations that are too big. The discovery of cold dark matter in some form or another in the lab would enable cosmologists to hone their ideas. There is a very healthy interplay going on.

Yet in all this optimism, it is very easy to lose sight of the one remarkable fact—that we are talking about the creation of the universe in a scientific way and that we hope to answer some of the questions by experiment.

Many have said that it is part of being human to have a curiosity about the world. The problem of creation has inspired humanity for centuries. Now we are starting to investigate these ideas scientifically rather than just through philosophy and theology. I doubt that science will ever provide the answers fully, but it must help to frame the philosophical discussion.

This is a most significant stage in the history of ideas.

Appendix 1

Nobel Prizes in physics

In the following list of Nobel Prize awards, contributions directly relevant to particle physics and cosmology are in bold type (this is of course a personal judgement).

Year	Name of winner	Citation
1901	Wilhelm Konrad Röntgen	Discovery of X-rays
1902	Hendrick Anton Lorentz Peter Zeeman	Effect of magnetic fields on light emitted from atoms
1903	Antoine Henri Becquerel Pierre Curie Marie Sklowdowska-Curie	Discovery of radioactivity Research into radioactivity
1904	Lord John William Strutt Rayleigh	Discovery of argon and measurements of densities of gases
1905	Philipp Eduard Anton von Lenard	Work on cathode rays
1906	**Sir Joseph John Thompson**	**Conduction of electricity by gases**
1907	Albert Abraham Michelson	Optical precision instruments and experiments carried out with them
1908	Gabriel Lippmann	For his method of producing colours photographically using interference
1909	Guglielmo Marconi Carl Ferdinand Braun	Development of wireless telegraphy

1910	Johannes Diderik van der Waals	Equations of state for gases and liquids
1911	Wilhelm Wein	Laws governing the radiation of heat
1912	Nils Gustaf Dalen	Invention of automatic regulators for lighthouse and buoy lamps
1913	Heike Kamerlingh Onnes	Properties of matter at low temperatures leading to the discovery of liquid helium
1914	Max von Laue	Discovery of X-ray diffraction by crystals
1915	William Henry Bragg William Lawrence Bragg	Analysis of crystal structure using X-rays
1916		Prize money allocated to the special fund of this prize section
1917	Charles Glover Barkla	Discovery of characteristic X-rays from elements
1918	**Max Planck**	**Discovery of energy quanta**
1919	Johannes Stark	Discovery of Doppler effect in canal rays and the splitting of spectral lines in magnetic fields
1920	Charles-Edouard Guillaume	Services to precision measurements in physics by his discovery of anomalies in nickel steel alloys
1921	Albert Einstein	Services to theoretical physics especially explanation of photoelectric effect
1922	Niels Bohr	Structure of atoms and radiation from them
1923	**Robert Andrew Millikan**	**Measurement of charge on electron and work on photoelectric effect**
1924	Karl Manne Georg Siegbahn	X-ray spectroscopy
1925	James Franck Gustav Hertz	Experimental investigation of energy levels within atoms

1926	Jean Baptiste Perrin	Work on discontinuous nature of matter especially discovery of sedimentation equilibrium
1927	**Arthur Holly Compton** **Charles Thompson Rees Wilson**	**Discovery of Compton effect invention of the cloud chamber**
1928	Owen Willans Richardson	Work on thermionic phenomena
1929	Prince Louis-Victor de Broglie	Discovery of the wave nature of electrons
1930	Sir Chandrasekhara Venkata Raman	Work on scattering of light
1931		Prize money allocated to the special fund of this prize section
1932	**Werner Heisenberg**	**Creation of quantum mechanics**
1933	**Erwin Schrödinger** **Paul Adrien Maurice Dirac**	**Discovery of new productive forms of atomic theory (quantum mechanics)**
1934		Prize money allocated to the special fund of this prize section
1935	**James Chadwick**	**Discovery of the neutron**
1936	**Victor Franz Hess** **Carl David Anderson**	**Discovery of cosmic rays** **Discovery of the positron**
1937	Clinton Joseph Davisson George Paget Thompson	Discovery of electron diffraction by crystals
1938	Enrico Fermi	Discovery of nuclear reactions brought about by slow neutrons (fission)
1939	**Ernest Orlando Lawrence**	**Invention of cyclotron**
1940– 1942		Prize money allocated to the special fund of this prize section
1943	Otto Stern	Discovery of the magnetic moment of the proton

1944	Isidor Isaac Rabi	Resonance recording of magnetic properties of nuclei
1945	Wolfgang Pauli	Discovery of the exclusion principle
1946	Percy Williams Bridgman	Invention of apparatus to produce extremely high pressures and discoveries made with it
1947	Sir Edward Victor Appleton	Investigation of the physics of the upper atmosphere
1948	**Patrick Maynard Stuart Blackett**	**Development of Wilson cloud chamber and discoveries made with it**
1949	**Hideki Yukawa**	**Prediction of mesons**
1950	**Cecil Frank Powell**	**Photographic method for recording particle tracks and discoveries made with it**
1951	Sir John Douglas Cockroft Ernest Thomas Sinton Walton	Transforming atomic nuclei with artificially accelerated atomic particles
1952	Felix Bloch Edward Mills Purcell	Development of new methods for nuclear magnetic precision measurements and discoveries made with this technique
1953	Frits Zernike	Invention of the phase contrast microscope
1954	Max Born	Fundamental research in quantum mechanics
	Walter Bothe	The coincidence method and discoveries made with it
1955	Willis Eugine Lamb	Discoveries concerning the fine structure of the hydrogen spectrum
	Polykarp Kusch	Precision measurement of the magnetic moment of the electron
1956	William Shockley John Bardeen Walter Houser Brattain	Discovery of transistor effect

1957	Chen Ning Yang Tsung Dao Lee	Prediction of parity non-conservation
1958	Pavel Aleksejevic Čerenkov Il'ja Michajlovic Frank Igor' Evan'evic Tamm	Discovery and interpretation of Čerenkov effect
1959	Emilio Gino Segrè Owen Chamberlin	Discovery of the antiproton
1960	Donald Arthur Glasser	Invention of the bubble chamber
1961	Robert Hofstader	Studies in the electron scattering of atomic nuclei
	Rudolf Ludwig Mössbauer	Research into the resonant absorption of γ rays
1962	Lev Davidovic Landau	Theory of liquid helium
1963	Eugeine P Wigner	Theory of atomic nucleus
	Maria Goeppert Meyer J Hans Jensen	Discovery of nuclear shell structure
1964	Charles H Townes Nikolai G Basov Alexander M Prochrov	Quantum electronics and masers/lasers
1965	Sin-itiro Tomonaga Julian Schwinger Richard Feynman	Work on quantum electrodynamics
1966	Alfred Kastler	Discovery of optical methods for studying Hertzian resonance in atoms
1967	Hans Albrecht Bethe	Contributions to the theory of energy production in stars
1968	Luis W Alvarez	Discovery of resonance particles and development of bubble chamber techniques
1969	Murray Gell-Mann	Discoveries concerning the classification of elementary particles

1970	Hannes Alfvén	Discoveries in magnetohydrodynamics
	Louis Néel	Discoveries in antiferromagnetism and ferrimagnetism
1971	Dennis Gabor	Invention of holography
1972	John Bardeen	Jointly developed theory of superconductivity
	Leon N Cooper	
	J Robert Schrieffer	
1973	Leo Esaki	Discovery of tunnelling in semiconductors
	Ivar Giaever	Discovery of tunnelling in superconductors
	Brian D Josephson	Theory of super-current tunnelling
1974	**Antony Hewish**	**Discovery of pulsars**
	Sir Martin Ryle	**pioneering work in radioastronomy**
1975	Aage Bohr	Discovery of the connection between collective motion and particle motion in atomic nuclei
	Ben Mottleson	
	James Rainwater	
1976	**Burton Richter**	**Independent discovery of J/ψ particle**
	Samuel Chao Chung Ting	
1977	Philip Warren Anderson	Theory of magnetic and disordered systems
	Nevill Frances Mott	
	John Hasbrouck Van Vleck	
1978	Peter L Kapitza	Basic inventions and discoveries in low temperature physics
1978	**Arno Penzias**	**Discovery of the cosmic microwave background radiation**
	Robert Woodrow Wilson	
1979	**Sheldon Lee Glashow**	**Unification of electromagnetic and weak forces**
	Abdus Salam	
	Steven Weinberg	
1980	**James Cronin**	**Discovery of K^0 CP violation**
	Val Fitch	
1981	Nicolaas Bloembergen	Contributions to the development of laser spectroscopy
	Arthur L Schalow	
	Kai M Siegbahn	Contribution to the development of high resolution electron spectroscopy

1982	Kenneth G Wilson	Theory of critical phenomena in phase transitions
1983	**William Fowler**	**Theoretical and experimental studies of nucleosynthesis of elements inside stars**
	Subrahmanyan Chandrasekhar	**Theoretical studies in the structure and evolution of stars**
1984	**Carlo Rubbia Simon van der Meer**	**Discovery of W and Z particles**
1985	Klaus von Klitzing	Discovery of quantum Hall effect
1986	Ernst Ruska Gerd Binning Heinrich Rohrer	Design of electron microscope Design of the scanning tunnelling microscope
1987	Georg Bednorz K A Müller	Discovery of high temperature superconductivity
1988	**Leon M Lederman Melvin Schwartz Jack Steinberger**	**Neutrino beam method and demonstration of two kinds of neutrino**
1989	Normal Ramsey	Separated field method of studying atomic transitions
	Hans Dehmelt Wolfgang Paul	Development of ion traps
1990	**Jerome Friedman Henry Kendall Richard Taylor**	**Research into deep inelastic scattering**
1991	Pierre-Gilles de Gennes	Mathematics of molecular behaviour in liquids near to solidification
1992	**Georges Charpak**	**Invention of multiwire proportional chamber**
1993	**Joseph Taylor Russell Hulse**	**Discovery of a binary pulsar and research into gravity based on this**
1994	Bertram Blockhouse Clifford Shull	Development of neutron spectroscopy development of neutron diffraction techniques

1995	**Fredrick Reines**	**Discovery of electron-neutrino**
	Martin Perl	**Discovery of tau-meson**
1996	David M Lee	Discovery of superfluid helium-3
	Douglas D Osheroff	
	C Richardson	
1997	Steven Chu	Development of methods to cool and
	Claude Cohen-Tannoudji	trap atoms with laser light
	William D Phillips	

Appendix 2

Glossary

amplitude
A quantity calculated in quantum mechanics in order to obtain the probability of an event. Amplitudes must be absolute squared to obtained probabilities. They do not add like ordinary numbers, a fact that explains some of the odd effects of quantum mechanics.

antimatter
Every particle of matter has an antimatter partner with the same mass. Other properties, such as electrical charge, baryon number and lepton number are reversed. If matter and antimatter particles meet they annihilate into energy.

atom
Atoms consist of a positively charged nucleus around which electrons move in complicated paths. The nucleus contains protons and neutrons and has the bulk of the mass of the atom. Nuclei are typically a hundred thousand times smaller than atoms. Atoms are the smallest chemically reacting entities.

baryon
A class of particle. Baryons are a subclass of hadrons, and so they are composed of quarks. All baryons are composed of three quarks. Antibaryons are composed of three antiquarks.

big bang
Current experimental evidence suggests that the universe originated in a period of incredibly high density and temperature called the big bang.

Since that time, space has been expanding and the universe has been cooling. The primary evidence for this comes from Hubble's law and the cosmic microwave background radiation.

black body radiation
Any object that is in thermal equilibrium with the electromagnetic radiation it is emitting produces a characteristic spectrum of wavelengths known as black body radiation. The spectrum depends only on the temperature of the object, not on the material it is made from. The cosmic microwave background radiation has a black body spectrum of temperature 2.78 K.

bottom quark
A member of the third generation of quarks. First discovered in 1977 in the upsilon meson (bottom, antibottom).

brown dwarf
A large gaseous object intermediate in size between the planet Jupiter and a small star. As such, it is not big enough to light up nuclear reactions and become a genuine star. Recent observations of brown dwarfs orbiting stars suggest that such objects may be quite common in the universe.

charm quark
The charm quark is a member of the second generation. Any hadron containing a charm quark is said to have the property *charm*.

color
Quarks have a fundamental property called color which comes in three types: red, blue and green. Color plays a similar role in the theory of the strong force as electric charge does in the theory of electromagnetism. Antiparticles have anticolor: antired, antiblue and antigreen. All hadrons must be colorless—a baryon must have one quark of each color, a meson has a color, anticolor pair (i.e. blue, antiblue, etc). Leptons do not possess color, and so do not feel the strong force.

conservation law
The universe contains a fixed amount of energy and although it is split between different types (kinetic, potential, etc) the sum total of all

the amounts remains the same for all time. This is conservation of energy. Charge is another conserved quantity. The total amount of charge (counting positive charges and negative charges) must remain the same. Other conservation laws are useful in particle physics such as conservation of baryon number and lepton numbers. Physicists are pleased to find conservation laws as they often express fundamental properties of forces.

cosmic rays
These are high energy particles that pour down from space. Many of them have their origin in the solar wind produced by the sun (high energy protons). Some come from other parts of our galaxy. Often these particles react with the atoms in the upper atmosphere producing pions. These pions decay into muons on their way down to earth. The majority of cosmic rays detected near to the surface are muons produced in this way.

coulomb
The fundamental unit of electrical charge. In particle physics the coulomb is too large a unit for convenience and the standard unit is set to be the same as the charge on the proton, written as $e = 1.6 \times 10^{-19}$ coulombs.

cyclotron
A type of particle accelerator. First designed by E O Lawrence. The device relies on the periodic 'kicking' of a charged particle as it moves in a circular path through a potential difference.

dark matter
Matter that does not interact with electromagnetic radiation and so cannot be astronomically observed directly. Dark matter is needed to explain how stars move inside galaxies. It has also been included in theories of galaxy formation. The inflationary theory implies that most of the universe is composed of dark matter. Hot dark matter is particles moving close to the speed of light. Cold dark matter particles are moving at speeds very much slower than light.

deuterium
An isotope of hydrogen in which there is one proton and one neutron in the nucleus.

down quark
A flavour of quark from the first generation.

eightfold way
The classification of hadrons devised by Murray Gell-Mann based on plotting diagrams of particle families by mass.

electromagnetism
A fundamental force of nature. Static electrical and magnetic forces are actually simple examples of a more general electromagnetic force that acts on charges and currents. In this book electromagnetism is used as a generic name for electrical and magnetic forces.

electron
A fundamental particle and the lightest of the massive particles in the lepton family.

electron-volt (eV)
A unit of energy. One electron-volt is the amount of energy an electron would gain if it were accelerated through a potential difference of one volt.

$1 \text{ eV} = 1.6 \times 10^{-19} \text{ J}$

1 keV is a thousand eV

1 MeV is a million eV

1 GeV is a thousand million eV

1 TeV is a million million eV.

electroweak force
In the 1970s it was shown that the weak force could be grouped into a single theory of particle interactions with the electromagnetic force. The resulting theory refers to the electroweak force. In the theory four fundamental fields couple and combine to produce the W^+, W^-, Z^0 and the γ.

elementary particle
A fundamental constituent of nature. Current theory suggests that quarks and leptons (and their antiparticles) are fundamental particles in that they are not composed of combinations of other smaller particles.

exchange particle
A particle that is formed from the interaction between two other particles. For example, a photon that passes between two electrons when they interact electromagnetically is an exchange particle. One of the most important successes of particle physics theory in recent times has been the general theory of exchange particles that developed from QED to give us QCD, electroweak unification and the various GUTs.

filaments
Filaments are structures in space composed of many galaxies and clusters of galaxies that seem to be aligned into long thin 'strings' or filaments.

Feynman diagram
Feynman diagrams are often used to represent different aspects of the processes that go on when particles interact via the fundamental forces. They were originally designed as a way of showing how the various mathematical terms in a calculation linked together.

flavour
A whimsical name for the different types of quark and lepton that exist. There are six flavours of quark (up, down, strange, top and bottom) and six flavours of lepton (electron, electron-neutrino, muon, muon-neutrino, tau and tau-neutrino).

generation
Quarks and leptons can be grouped into generations by the way they react to the weak force. There are three generations of lepton and three generations of quark.

gluon
The exchange particle responsible for the strong force between quarks. Gluons are to the strong field as photons are to the electromagnetic field.

hadron
A type of elementary particle composed of quarks. The strong force acts on the quarks in such a way as to ensure that quarks can never be found in isolation—they must always be bound into hadrons. Hadrons can be split into two sub-categories—the baryons (and antibaryons) and the mesons.

Higgs particle

The Higgs field is thought to exist and the interaction between this field and the various elementary particles is what gives rise to their mass. If the field exists then there should be excitations of the field called Higgs particles (as the excitations of the electromagnetic field are photons). As yet there is no direct experimental evidence for the Higgs. The idea of the Higgs field has proven useful in solving some of the problems of the standard model of cosmology—especially in the context of inflationary cosmology.

inflation

A recent theory that modifies the big bang evolution of the universe. The suggestion is that early in the universe's history it grew very rapidly in size (inflated). The mechanisms that drive this inflation would also have the effect of forcing the universe's density towards the critical value. Consequently, we expect the universe's density to be very close to the critical density now if inflation is true. The amount of visible matter in the universe is a lot less than this, so inflation requires dark matter to make up the rest of the density.

ion

An atom from which one or more electrons have been removed. The resulting particle is positively charged. The electrons which have been removed are also sometimes referred to as ions in this context.

kaon

A strange meson. There are four kaons K^+, K^-, K^0 and $\bar{K}^0$.

kelvin scale

The absolute temperature (or kelvin) scale is a measurement of temperature that is not based on any human definition. The centigrade and Fahrenheit scales are based on arbitrary (but sensible) definitions that we have chosen. Absolute temperature is dictated by the properties of gases. The lowest possible temperature is absolute zero (zero kelvin) which corresponds to $-273\ {}^\circ$C.

Lagrangian

The difference between the kinetic energy of a particle and its potential energy at any point in space and time is the Lagrangian of the particle. Lagrangians are well known in classical mechanics and Feynman extended their use to become the basis of his version of quantum mechanics.

lepton
A class of elementary particle. Leptons can interact via all the fundamental forces except the strong force. There are six flavours of lepton to match the six flavours of quark.

lifetime
Unstable particles have a lifetime which is the average time taken for particles of that type to decay. There is no way to determine the moment at which an individual particle will decay as the process is inherently random.

mass shell
Sometimes a particle can be emitted with a mass different than normal. When this happens the particle is said to be 'off mass shell'. This only happens as an intermediate step in an interaction such as those illustrated in Feynman diagrams—the exchange particle is off mass shell. Such particles are sometimes called virtual particles as they can not be directly observed.

meson
A type of particle composed of a quark and an antiquark bound together. As they contain quarks, mesons are classed as hadrons.

molecule
A structure composed of a group of atoms bound together. Molecules can be quite simple, such as salt (NaCl) or water (H_2O) or very complex containing millions of atoms such as DNA.

muon
A lepton from the second generation. Muons are effectively heavy electrons differing only in that they have muon-number rather than electron-number. Muons are abundant in the cosmic rays that reach ground level.

neutrino
Every massive lepton has an (apparently) massless neutrino with it in a lepton generation. Neutrinos have no electrical charge so they can

only interact via the weak force. This makes them very difficult to experiment with as they interact so infrequently. Electron-neutrinos were predicted by Pauli in the 1930s as a way of helping to explain the energy released in nuclear β decay. However, as neutrinos are very weakly interacting, the discovery of the electron-neutrino had to wait until the 1950s when large numbers of them were available from nuclear reactors.

nucleon
General name for any particle contained within a nucleus, i.e. protons and neutrons are nucleons.

nucleus
Part of the atom. The nucleus contains protons and neutrons and so carries a positive electric charge. The electrons of the atom orbit the nucleus in complicated paths. Nuclei are very much smaller than the volume of the atom itself (10^{25} times smaller) yet contain over 99% of the mass of the atom. Nuclei are extremely dense.

phase
Any quantity that goes through a cyclic motion has a phase which marks the point in the cycle—e.g. the phases of the moon. Phase is normally specified as an angle, with a phase of 2π radians indicating that the cycle is complete. Quantum mechanical amplitudes have both size and phase.

photon
Photons are 'particles' of light. They are the elementary excitations of the electromagnetic field. Photons have zero charge and no mass. The exchange of virtual photons between charged particles gives rise to the electromagnetic force.

pion
The three pions are the lightest particles in the meson class. They are composed of u and d type quarks and antiquarks. Pions are very easily produced in reactions between hadrons.

plasma
A very hot gas in which all the atoms have been ionized.

positron
The antiparticle of the electron.

proton
The lightest particle in the baryon class. Protons are composed of two u quarks and a d quark.

QED, QCD
The theory of the electromagnetic force is known as quantum electrodynamics (QED) and it is one of the most accurate theories that has ever been produced. Its predictions have been experimentally checked to a very high degree of accuracy. Richard Feynman invented Feynman diagrams to make the calculations of QED easier to organize. QED has been used as the model for all theories of the fundamental forces, in particular quantum chromodynamics (QCD) is the theory of the strong force and it was created by making a simple extension to the ideas of QED.

quark
A fundamental particle of nature. Quarks experience all the fundamental forces. The strong force acts to bind them into composite particles called hadrons. Three quarks made a baryon, three antiquarks make an antibaryon and a quark paired with an antiquark makes a meson. These are the only combinations allowed.

quantum mechanics
The modern theory of matter that has replaced Newton's mechanics. Quantum theory has many odd consequences including the inability to distinguish particles and having to include all possible paths of a particle when computing an amplitude.

radian (rad)
A measurement of angle based on the circumference of a circle of radius 1. $360°$ is the same as 2π radians, $90°$ is $\pi/2$ radians, etc.

radioactivity
The nuclei of some atoms are unstable. The stability of a nucleus depends critically on the number of protons and neutrons within it. Some isotopes have the wrong number of neutrons for the number of protons. This makes the nucleus unstable. The nucleus can become

more stable by getting rid of excess energy in the form of a gamma ray, or by emitting an alpha particle (two protons and two neutrons bound together) or a beta particle (an electron), or a positron. In the latter two cases, either a neutron has decayed into a proton within the nucleus by emitting an electron or a proton has converted into a neutron by emitting a positron.

recombination

Recombination is the name given to the epoch in the universe's history at which electrons combined with the nuclei of hydrogen and helium to form neutral atoms. It is a misnomer in that the electrons had never before been permanently bound to atoms. Before this time the energy in the photons flooding the universe was greater than the energy required to split up atoms, so the electrons were always blasted away. After this epoch there are essentially no free electrical charges in the universe and photons are much less likely to interact with matter.

spin

A fundamental property of particles. Spin has no classical equivalent— it is a quantum mechanical property. The closest classical idea is to imagine that particles spin like tops. However, a top can spin at any rate, whereas particles can only spin at a rate that is a multiple of $h/2\pi$. Fermions spin at odd multiples of $h/4\pi$ (e.g. quarks and leptons which spin at $h/4\pi$). Bosons (e.g. photons, gluons, etc) spin at $h/2\pi$ or h/π.

strangeness

Any hadron that contains a strange quark is said to have the property strangeness. This property can be treated a little like lepton number as it is conserved in all reactions that do not involve the weak force. The strangeness of a hadron can be -1 (one strange quark), -2 (two s quarks), -3 (three s quarks), $+1$ (one antistrange quark), 0 (either no strange quarks or an s$\bar{\text{s}}$ combination) etc.

strong force

A fundamental force of nature. Quarks feel the strong force as they have the property color. The strong force is mediated by exchange particles called gluons. Gluons also carry color and so are subject to the force themselves. This simple difference between the strong force and the electromagnetic (photons do not have charge and so do not feel the force that they mediate) makes a huge difference in the properties of the force. Whereas the electromagnetic force decreases with distance, the strong force gets stronger. We suspect that because of this the strong force will never allow a quark to be seen in isolation.

superconductors
Materials that have zero resistance to electric current when they are cooled to a low temperature (typically a few kelvin). There is a great deal of research going on at the moment to try to design a material that will be a superconductor at something close to room temperature.

supersymmetry
A theory that relates fermions to bosons. The theory predicts that every fermion particle should have a boson equivalent (e.g. a quark will have a squark) and that every boson should have an equivalent fermion (e.g. photon and photino). There have been some recent hints that supersymmetric particles exist, but as yet none of them have been directly observed.

synchrotron
The first type of particle accelerator was the synchrotron. It relies on the charged particle circulating round in a circular cavity. The cavity is divided into two halves (to form D segments). As the particle crosses from one half to the other it gets a kick from an electric field. This kick can be synchronized (hence the name) to the circulation of the particle, so that it can be repeatedly accelerated.

tau
The heaviest of the lepton family of particles.

thermal equilibrium
A system is in thermal equilibrium when there is no net transfer of energy from one part to another.

top quark
The most recent quark to be discovered. It is very much more massive than any of the others and its discovery took a long time as particle accelerators did not have enough energy to create it.

unification
Particle physicists hope that at some point in the future they will be able

to combine all the theories of the electroweak, strong and gravitational forces into one unified theory. When this happens, the different forces will be seen as different aspects of the one underlying force. There have been some tentative hints that this may be possible and many partially developed theories. Grand Unified Theories (GUTs) combine the electroweak and strong forces, but we have no experimental evidence as yet to decide which one is right. Theories of Everything (TOEs) are more speculative as they include gravity as well. The best candidate TOE at the moment is superstring theory, although this is not fully developed mathematically as yet.

up quark
A flavour of quark from the first generation. Up and down quarks are the most commonly found quarks in nature as they are the components of protons and neutrons.

virtual particle
A general name for a particle that is off mass shell.

voids
Gigantic regions in space in which there are far fewer galaxies than in the surrounding space.

W and Z particles
These are the mediators of the weak force. Theories of the weak force contained the W^+ and W^-, but the electroweak theory of Salam, Glashow and Weinberg suggested that the combination of weak and electromagnetic forces required the Z as well if it was to work. The first success of this theory was the discovery of the neutral current—an interaction mediated by the Z.

xenon
An element. Xenon is a very unreactive gas and a heavy atom. It has 54 protons in its nucleus. Xenon gas was used in the recent antihydrogen-producing experiment.

Appendix 3

Particle data tables

The quarks

Name	Symbol	Charge (+e)	Mass (GeV/c^2)	Stable	Date	Discovererd/predicted by	How/where
up	u	+2/3	0.33	yes	1964	Gell-Mann Zweig	quark model
down	d	−1/3	~0.33	no	1964	Gell-Mann Zweig	quark model
charm	c	+2/3	1.58	no	1974	Richter *et al* Ting *et al*	SIAC BNL
strange	s	−1/3	0.47	no	1964	Rochester and Butler	strange particles
top	t	+2/3	180	no	1995	CDF collaboration	Fermilab
bottom	b	−1/3	4.58	no	1977	Lederman *et al*	Fermilab

The leptons

Name	Symbol	Charge (+e)	Mass (GeV/c^2)	Stable/ lifetime (s)	Date	Discoverers	How/where
electron	e^-	-1	5.31×10^{-4}	yes	1897	J J Thomson	cathode ray tube
electron-neutrino	ν_e	0	0	yes	1956	E Cowan F Reines	reactor
muon	μ^-	-1	0.106	2×10^{-6}	1937	S Neddermeyer C Anderson	cosmic rays
muon-neutrino	ν_μ	0	0	yes	1962	M Schwartz *et al*	BNL
tau	τ^-	-1	1.78	3×10^{-13}	1975	M Perl *et al*	SLAC
tau-neutrino	ν_τ	0	0	yes	1978	indirect evidence	—

Some baryons

Name	Symbol	Charge (+e)	Mass (GeV/c^2)	Lifetime (s)	Quarks	Strangeness	Date of discovery
proton	p	+1	0.938	$> 10^{39}$	uud	0	1911
neutron	n	0	0.940	900	udd	0	1932
lambda	Λ	0	1.115	2.6×10^{-10}	uds	−1	1947
sigma plus	Σ^+	+1	1.189	0.8×10^{-10}	uus	−1	1953
sigma minus	Σ^-	−1	1.197	1.5×10^{-10}	dds	−1	1953
sigma zero	Σ^0	0	1.192	6×10^{-20}	uds	−1	1956
xi minus	Ξ^-	−1	1.321	1.6×10^{-10}	dss	−2	1952
xi zero	Ξ^0	0	1.315	3×10^{-10}	uss	−2	1952
omega minus	Ω^-	−1	1.672	0.8×10^{-10}	sss	−3	1964
lamda c	Λ_c	+1	2.28	2.3×10^{-13}	udc	+1 charm	1975

Some mesons

Name	Symbol	Charge (+e)	Mass (GeV/c^2)	Lifetime (s)	Quarks	Strangeness	Date of discovery
pi zero	π^0	0	0.135	0.8×10^{-16}	$u\bar{u}/d\bar{d}$	0	1949
pi plus	π^+	+1	0.140	2.6×10^{-8}	$u\bar{d}$	0	1947
pi minus	π^-	−1	1.115	2.6×10^{-10}	$d\bar{u}$	0	1947
K zero	K^0	0	0.498	*	$d\bar{s}$	+1	1947
K plus	K^+	+1	0.494	1.2×10^{-8}	$u\bar{s}$	+1	1947
K minus	K^-	−1	0.494	1.2×10^{-8}	$s\bar{u}$	−1	1947

* The K^0 and the $\bar{K}^0$ are capable of 'mixing' as the weak force can change one into the other—this makes it impossible to measure their lifetimes separately.

Some more mesons

Name	Symbol	Charge (+e)	Mass (GeV/c^2)	Lifetime (s)	Quarks	Charm	Date of discovery
psi	ψ	0	3.1	10^{-20}	$c\bar{c}$	0	1974
D zero	D^0	0	1.86	4.3×10^{-13}	$c\bar{u}$	+1	1976
D plus	D^+	+1	1.87	9.2×10^{-13}	$c\bar{d}$	+1	1976

Appendix 4

Further reading

There are many excellent books on the market. The following is a list of personal favourites and reference sources used in writing this book.

QED The Strange Theory of Light and Matter
R P Feynman, Penguin Science
(Feynman's explanation of quantum mechanics to a lay audience)

The Character of Physical Law
R P Feynman, Penguin Science
(This is a book about the laws of nature, but it has a good section on quantum mechanics)

The meaning of quantum mechanics is discussed in:
The Mystery of the Quantum World
E Squires, Adam Hilger

Particle and fundamental physics including some relativity is covered in:
To Acknowledge the Wonder
E Squires, Adam Hilger

For people who want to push things a little further there are:
Superstrings and the Search for the Theory of Everything
F David Peat, Abacus

Gauge Theories in Particle Physics
I J R Aitcheson and A J G Hey, Adam Hilger
(The text from which I learned much of my particle physics)

Introduction to Particle Physics
D Griffiths, Wiley
(An excellent introductory text that is very readable)

Essential Relativity
W Rindler, Springer
(In my opinion the best book on relativity—with an introduction to cosmology)

Relativity: An Introduction to Space-Time Physics
S Adams, Taylor and Francis (a good friend of mine—so read it!)

The Ideas of Particle Physics
J E Dodd, Cambridge University Press

Two wonderful books on astronomy, cosmology and the people who work in those fields are:
First Light, the Search for the Edge of the Universe
R Preston, Abacus

Lonely Hearts of the Cosmos, the Quest for the Secret of the Universe
D Overbye, Picador

Good popular books on cosmology, dark matter and such are:
The Inflationary Universe
A H Guth, Jonathan Cape

Wrinkles in Time
G Smoot and K Davidson, Abacus

The Shadows of Creation
M Riordan and D Schramm, Oxford University Press

The Stuff of the Universe
J Gribbin and M Rees, Penguin Science

The Primeval Universe
J V Narlikar, Oxford University Press
(More 'meaty' than the others—it includes some maths)

Bubbles, Voids and Bumps in Time: The New Cosmology
J Cornell (ed), Cambridge University Press
(Experts in various areas contribute chapters)

Index